AF520140

The Role of Science Popularization in Science, Technology and Innovation Policy

The Role of Science Popularization in Science, Technology and Innovation Policy

Edited by

Dr. Akram Ghadimi
Director of Science Popularisation Department
NRISP, Tehran, Iran

Mr. Hasan Jawaid Khan
Chief Scientist and Head
Popular Science Division, CSIR-NISCAIR, New Delhi, India

Editorial Support Team from NRISP, Iran
(Sahar Kousari, Mojgan Samandar
Ali Eshtehardi and Zohair Sabaghpour)

A Publication of the Centre for Science & Technology of the Non-Aligned and Other Developing Countries (NAM S&T Centre)

ALLIED PUBLISHERS PVT. LTD.

New Delhi • Mumbai • Kolkata • Chennai • Bangalore • Hyderabad

ALLIED PUBLISHERS PRIVATE LIMITED

1/13-14 Asaf Ali Road, **New Delhi**–110002
Ph.: 011-23239001 • E-mail: delhi.books@alliedpublishers.com

17 Chittaranjan Avenue, **Kolkata**–700072
Ph.: 033-22129618 • E-mail: cal.books@alliedpublishers.com

15 J.N. Heredia Marg, Ballard Estate, **Mumbai**–400001
Ph.: 022-42126969 • E-mail: mumbai.books@alliedpublishers.com

No. 25/10, Commander-in-Chief Road, Ethiraj Lane (Next to Post Office)
Egmore, **Chennai**–600008
Ph.: 044-28223938 • E-mail: chennai.books@alliedpublishers.com

P.B. No. 9932, No. 15, 3rd Floor (Next to Vijaya Bank), 5th Cross, Gandhinagar,
Karnataka, **Bangalore**–560009, Ph.: 080-41530285 / 22386239
• E-mail: bngl.journals@alliedpublishers.com / apsabng@airtelmail.in

Sri Jayalakshmi Nilayam, No. 3-4-510, 3rd Floor (Above More Super Market)
Barkatpura, **Hyderabad**–500027
Ph.: 040-27551811, 040-27551812 • E-mail: hyd.books@alliedpublishers.com

Website: www.alliedpublishers.com

A Publication of the Centre for Science & Technology of the
Non-Aligned and Other Developing Countries (NAM S&T Centre), New Delhi

© 2021, All rights reserved with the NAM S&T Centre

ISBN: 978-93-89934-22-9

Contributing Staff Editor: Gloria Susan Cherian, Research Associate

No part of the material protected by this copyright notice may be reproduced or utilized in any form or by any means, electronic or mechanical including photocopying, recording or by any information storage and retrieval system, without prior written permission from the copyright owners. The views expressed in this volume are of the individual contributors, editor or author and do not represent the view point of the Centre.

Published by Sunil Sachdev and printed by Ravi Sachdev at Allied Publishers Pvt. Ltd.,
1/13-14 Asaf Ali Road, New Delhi–110002

science & innovation
Department:
Science and Innovation
REPUBLIC OF SOUTH AFRICA

Private Bag X894, Pretoria, 0001 Tel.: +27 (0)12 843 6300 Fax: +27 (0)12 349 1030 www.dst.gov.za

Foreword

The role of Science and Technology (S&T) is getting recognized globally as an engine of development. However, for S&T to translate into improved lifestyle, wealth and employment creation, it must be geared towards innovation: the process of bringing new goods and services to the market.

Production units in all developed economies are been driven by the utilisation of scientific and technological knowledge. The need of the hour, therefore, is to put in place explicit strategies, mechanisms and institutions to promote the use of scientific knowledge for development. This further necessitates the formulation and reviewing of STI policies in driving sustainable development.

Ergo, I am pleased to learn that the Centre for Science and Technology of the Non-Aligned and Other Developing Countries (NAM S&T Centre) had organised an ***International Workshop on "Policy Making in Popularisation of Science, Technology and Innovation (STI)"***, in Tehran, Iran during 17–19 June, 2019 in collaboration with the National Research Institute for Science Policy (NRISP), Tehran. The Workshop was primarily oriented to provide a platform to scientists, technologists, researchers, academicians, industry people, policy makers and other concerned professionals working in the field of STI to deliberate and discuss the need to develop policies and strategies that favour the popularisation of science, technology and innovation in developing countries.

As a follow up, the NAM S&T Centre has brought out this book which comprises of 14 scientific papers contributed by experts and professionals from 9 different countries namely; Bhutan, Cuba, India, Iran, Mauritius, Nigeria, Sri Lanka, Togo and Zimbabwe, giving insights on developing various methodologies for 'Policy Making' in popularisation of STI, and in comprehending appropriate policy framework to encourage such policies and its diffusion. The volume has been edited profusely by Dr. Akram Ghadimi, Director, Science Popularisation Department, NRISP; Mr Hasan Jawaid Khan, Chief Scientist, CSIR-NISCAIR; and Dr Zohair Sabaghpour Azarian, Assistant Professor, NRISP.

I congratulate the NAM S&T Centre for bringing out this valuable publication which will be of immense use to scientists, researchers and others who are actively engaged in popularisation of STI in the developing countries.

Daan du Toit
Deputy Director-General,
International Cooperation and Resources,
Department of Science and Innovation,
Pretoria, South Africa

Introduction

One of the major challenges facing developing countries is to make *Science, Technology and Innovation* (STI) an essential part of the culture of the people. It is imperative for these countries to develop policies and strategies that favour the popularization of Science and Technology. Despite the wide variety in economic and social conditions, every country must make efforts to generate and nurture an endogenous S&T culture.

The scope of spreading knowledge in Science and Technology must be broad based to integrate formal education with informal efforts. It is also essential to make S&T knowledge available to the public at large, so as to inculcate basic understanding and create a general awareness about Science and Technology.

Science popularisation implies to bring science to the general public to disseminate scientific knowledge and to foster a scientific way of thinking among people. In other words, science popularization refers to public understanding of science. One of the very effective means of science popularisation is to make young students interested in science and careers in science by exposing them to an easily accessible non-formal science learning ambience where they can have the opportunity to practice science with their own hands, learn to put questions and discover the scientific principles involved, all by themselves.

Currently, science popularisation is comparatively a younger field in most of the developing countries and is still striving to carve out its rightful place in the domain of science. Popularisation of science includes a sufficiently broad collective process which involves research institutions, universities, governmental and non-governmental agencies that are working to communicate Science, Technology and Innovation to the masses, and to foster scientific and technological temper among people so that they are empowered to take considered decisions in their day to day life or on issues of national importance.

The developing countries must work together to formulate and implement a common policy with the objective of disseminating and popularising science, inculcating scientific temper and eradicating scientific illiteracy among their people.

In order to deliberate on various issues related to the need for appropriate policy supports to encourage **"Popularization of Science"** in the *Non-aligned and Other Developing Countries*, the *Centre for Science and Technology of the Non-Aligned and Other Developing Countries (NAM S&T Centre), New Delhi*, in partnership with the *Science Popularisation Department of the National Research Institute for Science Policy (NRISP), Tehran, Iran* organised an International Workshop on **'Policy Making in Popularisation of Science, Technology and Innovation** (STI)' during 17–19 June 2019. The three-day Workshop brought together **96** scientists, experts and professionals from **16** developing countries. As a follow up of the Tehran Workshop, the present book titled **"The Role of Science Popularization in Science, Technology & Innovation Policy"** is being published by the NAM S&T Centre which includes 14 papers contributed by the participants and other professionals and experts from nine developing countries namely: Bhutan, Cuba, India, Iran, Mauritius, Nigeria, Sri Lanka, Togo and Zimbabwe.

I highly appreciate the time and efforts put in place by the Editors of this book, Dr. Akram Ghadimi, Associate Professor of Science Popularization Department in the National Research Institute for Science Policy (NRISP), Tehran, Iran; Mr. Hasan Jawaid Khan, Chief Scientist, CSIR-National Institute of Science Communication and Information Resources (CSIR-NISCAIR), New Delhi, India, and the Scientific Editorial Committee from Iran including Dr. Sahar Kousari, Dr. Mojgan Samandar Ali Eshtehardi and Dr. Zohair Sabaghpour for the technical editing of the papers.

I take this opportunity to express my deepest gratitude to Mr. Daan du Toit, Deputy Director General: International Cooperation and Resources, Department of Science and Innovation (DSI), Pretoria, South Africa for writing the *"Foreword"* for this book in spite of his very busy schedule.

I also acknowledge the interest and valuable efforts of the entire team of the NAM S&T Centre, and especially thank Mr. M. Bandyopadhyay, Senior Adviser for his guidance and supervision; Ms. Gloria Susan Cherian, Research Associate for coordinating the entire publication process, proof reading of the manuscripts, and liaising with the authors of the papers; and Mr. Pankaj Buttan, Data

Processing Manager for taking all necessary steps including compiling, designing and formatting the book into its present form.

We are also thankful to Mr. Sharad Gupta, Publishing Consultant and Mr. Jagdish Singh from the Publication Department, Allied Publishers, New Delhi for their significant efforts in bringing out this valuable publication.

I am sure this book will be of great significance and immense use to all the actors who weave the threads of *"Popularisation of Science, Technology and Innovation"* together including scientists, researchers, technologists, communicators, journalists, students and all other stakeholders.

(Amitava Bandopadhyay, Ph.D.)
Director General
NAM S&T Centre
New Delhi, India

Preface

Today, the increasing development of Science, Technology, and Innovation has led to further expansion and deepening of related studies and discussions. However, the long-term success of science and technology policies and policymaking requires predicting policy outcomes, effectively uplifting science popularization, and regular feedback based on a powerful and reliable monitoring system of society.

The popularization of science and technology policies and its achievements, and then receiving feedback on the justification and rationality of policies, play a major role in providing valuable knowledge to support decision-making processes at all levels, improving and developing the capacity of individuals, networks, as well as ensuring and improving quality in all areas of Science, Technology, and Innovation.

In this book, to achieve a deeper understanding of the concept of science popularization, the emphasis is on the concept of science popularization based on the expert definition on historical, philosophical and theoretical contexts of formation and development of this concept and the place of science popularization in the policymaking cycle.

The promotion of science is not the only step in the policy-making process, but is an intermediary for communicating the various steps of this process and interacting with all of them. The result of this communication and systematic interaction is that it will support decision-making and policy-making on a variety of topics, including science, technology, and innovation. In general, science popularization policies seek to structure, integrate, optimize, and regulate the various steps of the policy cycle from scheduling to the emergence of consequences and effects.

While Nations around the world recognize the significance of taking science to the people, they have come to define the nature and function of popularization of science in the context of their own national goals and aspirations and the cultural milieu. But science and technology policy makers the world over have remained steadfast in their belief that popularization of science among various stakeholders holds the key to widespread acceptance of STI policies. This is amply displayed by the incorporation of science popularization, in its various manifestations, in science policies in many countries.

There is also an increasing realization that science, technology, and innovation dissemination for it to be truly meaningful cannot continue as a one-way channel. Rather science popularization and dissemination of scientific knowledge needs to take on a more inclusive and interactive nature whereby potentially significant inputs from various stakeholders in the society can also be woven into the science, technology, and innovation policies.

This book aims to provide a *MetaFrame* for better understanding of science, technology, and innovation policies by explaining the position of science, technology, and innovation dissemination in the policy-making cycle as one of the dimensions of this process. In different chapters of this book, readers will find interesting and practical material on the theoretical foundations, definitions, patterns, methods, processes, and systems of science, technology policies, and innovation popularization and dissemination.

The Editors would like to take this opportunity to acknowledge the contributions made by the Members of the Scientific Editorial Committee especially Dr. Sahar Kousari, Dr. Mojgan Samandar Ali Eshtehardi and Dr. Zohair Sabaghpour in regard to the preparation and compilation of this book.

We hope that this book being brought out by the NAM S&T Centre, New Delhi benefits the readers by providing useful and relevant information on the subject for scientists, technologists, policy makers and other stakeholders in the developing world.

Editors

Dr. Akram Ghadimi

Mr. Hasan Jawaid Khan

Contents

Section 3: STI Popularization Tools

Section 4: STI Popularization Case Studies

Tehran Resolution

Section 1

STI Popularization Policy Making

Popularization of Science, Technology and Innovation in Cuba

Héctor Arias Martín
Ministry of Science, Technology and Environment of Cuba
E-mail: hector@citma.gob.cu

ABSTRACT: In order to describe the current state of the popularization of science, technology and innovation in Cuba, the National System of Science, Technology and Innovation is briefly presented along with the main actions, organizations and general conditions for the dissemination of the scientific and technological knowledge. The methods used were documentary review and interviews with officials of great experience. It is concluded that since the 1960s many institutions have been created that have among their functions the development and dissemination of scientific knowledge and its applications. There are opportunities to approach science for all audiences. In spite of all this, the scientific culture of the general population and the interaction between science, technology and innovation entities, universities and companies are still considered insufficient. The promotion of science and its application for the development of the country is a priority of the government.

Keywords: Popularization, Science, Technology, Cuba, Innovation, Development, Policies.

1. INTRODUCTION

The access of the entire population to science and technology has been a purpose in constant compliance and development in Cuba since 1959. This political will was expressed in the speech of Fidel Castro Ruz on January 15, 1960: "*The future of our Homeland must necessarily be a future of men of science, it has to be a future of men of thought ...*". Numerous scientific institutions have been created since then. Different actions are projected and executed so that diverse publics can access and pay tribute to science, technology and innovation.

Cuba has a System of Science, Technology and Innovation (SCTI). This System, described on the CITMA (Ministry of Science,

Technology and Environment) website, "covers all branches of socio-economic and cultural development in the country", includes 222 Science, Technology and Innovation entities linked to all sectors, 52 universities and other organizations. The whole society participates in the System; legal persons, whether state or private entities and natural persons. The mission of the Science, Technology and Innovation System is: "To promote the generation, assimilation and application of knowledge and technologies; stimulate and encourage learning and innovation in all spheres of economic and social life in the country, in order to contribute to sustainable development, improve the quality of life in our society and strengthen cultural identity, security and defense of the nation and the preservation of its independence."

Currently science, technology and innovation are elements to prioritize for the development of the country. This is reflected in the public interventions of leaders and new policies and legal standards, including the Constitution of the Republic.

This is a scenario in which the promotion of science among the population and especially in the economic and productive sectors, and among entrepreneurs becomes more important. Strategies and well-targeted actions of popularization of science, technology and innovation could contribute to the country's development objectives.

2. MATERIALS AND METHODS

The main methods were documentary review and interview with experts. Work reports were reviewed: results of the Third National Survey of Innovation; the survey of science, technology and innovation indicators in 2018; legal norms; communications of the Director of Science, Technology and Innovation of CITMA on the new policies; websites of CITMA, of the Cuban Academy of Sciences, of the Science and Technology Forum, and of the main national media. CITMA officials with extensive experience were interviewed.

3. RESULTS AND DISCUSSION

Since the beginning of the Cuban Revolution, the government has had among its priorities the education of the population and the training of highly qualified professionals.

One of the main bases for this objective is access to all levels of education, from pre-school to post-graduate teaching and continuous learning. The Literacy Campaign is a milestone in national history, followed by the generalization of primary and secondary education.

The creation of several research centers mainly associated with Agriculture, Medicine and Biotechnology, promoted the development of Cuban science in general. In addition to scientific institutions, organizations focused on technical solutions were also created in companies and industries: the Science and Technology Forum, the National Association of Innovators and Rationalizers and the Youth Technical Brigades. These organizations have a more political and massive character and promote the participation of students and workers who receive social recognition and remuneration for their innovations.

The Law of Associations led to the foundation of multiple scientific or technical associations, which pursue with their work to contribute to the development of research and the application of the achievements of science and technology.

The emergence of new universities in all the provinces, the faculties of Distance Education and the Courses for Workers, allowed greater access to Higher Education. Then the scope was increased with the creation of municipal university centers.

At present we have programs and systematic actions, which can be enriched, but offer different age groups an exposure to science and technology: in primary and secondary schools the circles of interest on various scientific and technological subjects; science festivals with a majority participation of adolescents and young people; the university chairs of the Elderly; the Open Door Days of the Cuban Academy of Sciences and other scientific and educational institutions, and the University for All on television.

There are spaces for scientific dissemination in the flat press, among them the supplement *En Red*, of the *Juventud Rebelde* newspaper, and the *Juventud Técnica* magazine; in radio stations and national television channels, the *Antena* and *Passage to the unknown* programs stand out, also in radio stations and territorial telecenters. The agreement signed between the CITMA and the Cuban Institute of Radio and Television should contribute to the emergence of new programs and improve the transmission schedules. The Scientific Journalism Awards are delivered annually. There is a Circle of Scientific Journalism with representation in each province with which seminars and other training actions are organized every year.

More than 230 serial scientific-technological publications have been registered. The most represented subjects are: Medical Sciences, Pedagogy, Agricultural Sciences, Technological Sciences, Social Sciences and Humanities and Life Sciences. 68% of publications are electronic, which allows them to be available to a larger audience at a lower cost. It would be worth evaluating the sciences or specific topics that have more or less publications and the correspondence of this with the development projections.

All these conditions show the interest and work of the State for the development of science, its application and availability for the entire population. In spite of this, a specific policy for the popularization of science, technology and innovation has not been approved.

The investment indicators in science, technology and innovation activities show a significant increase in business financing, which suggests greater awareness among entrepreneurs about the importance of science and technology for business success. However, the results of the National Innovation Survey confirm the need to encourage links between companies in the productive and service sectors and science entities and universities.

Precisely this connection between science, technology and innovation and the production of goods and services is one of the priorities for the economic and social development of the country. This is contemplated in the Constitution of the Republic and in the recently approved policies.

The Constitution of the Republic includes several references to science and technology. It is established since 1976 and is ratified in the new Constitution approved on April 10, 2019, that the creative and investigative activity in science is free; the State promotes the advancement of science, technology and innovation as essential elements for economic and social development; the State guides, encourages and promotes education, science and culture in all its manifestations.

Recently, new policies and legal provisions were approved for science and technology parks, high-tech companies and interface companies.

This will promote the development of science, technology and innovation in the business sector and the link between companies and universities, promoting the production and application of science and technology. Press conferences, presentations in academic, scientific and governmental institutions in all provinces, debates in the media and other activities for the dissemination of these new policies and legal provisions have been developed.

Cuba developed the Third National Survey of Innovation in the business sector and the realization of a survey of public perception of science has been conceived for 2020. The results of these surveys will be useful in the design of strategies to encourage and guide Science, Technology and Innovation activities for the development of the country.

4. CONCLUSIONS

In Cuba, various organizations work for the development of science and its contribution to the progress of the country. Universal access to all levels of education, from pre-school, including special education, to post-graduate teaching and continuous learning, allows the instruction of people with dissimilar conditions.

A general and integrating conception favors the development of Science, Technology and Innovation activities in most sectors: Agriculture, Art, Communication, Construction, Economy, Education, Industry, Health, and Transport. The whole society participates in the National System of Science, Technology and Innovation.

The elaboration and approval of a policy specifically oriented to the Popularization of Science, Technology and Innovation could contribute to the articulation of all the actions and resources devoted to disseminating science and technology. With a successful policy of popularization of STI, we would be closer to being a country of men of science, of men of thought.

ACKNOWLEDGEMENTS

I appreciate the incentive and the collaboration of the Directorate of Science, Technology and Innovation, particularly Dr. Jesús Chía Garzón, to MSc. Maria Luisa Zamora Rodríguez and MSc. Armando Rodríguez Batista.

REFERENCES

[1] Castro, F. (1960). Speech delivered at the ceremony celebrated by the Speleological Society of Cuba, at the Academy of Sciences, on January 15, 1960, Havana, Cuba; http://www.fidelcastro.cu/es/discursos/discurso-pronunciado-en-el-acto-celebrado-por-la-sociedad-espeleologica-de-cuba-en-la (Accessed May 27, 2019).

[2] Ministry of Science, Technology and Environment, Description of Science, Technology and Innovation System, Havana, Cuba; http://www.citma.gob.cu/ciencia-4 (Accessed May 27, 2019).

[3] Rodríguez, A. (2019). Presentation, The Policies of Science, Technology and Innovation. Summary, Havana, Cuba.

Science, Technology and Innovation Policies in India—Thrust on Communication and Public Engagement

Hasan Jawaid Khan

CSIR-National Institute of Science Communication and Information Resources (CSIR-NISCAIR), Dr. K.S. Krishnan Marg, New Delhi–110012, India

E-mail: hjk@niscair.res.in

ABSTRACT: Science popularization and spreading scientific temper in culturally diverse societies is a complex undertaking. A complex set of linkages between science and a heterogeneous mass of population has to be understood and navigated. Communication and popularization of science among such heterogeneous masses therefore always presents a challenge.

Ever since India's independence, development of science and technology has gone hand in hand with science communication and popularization and inculcation of scientific temper among the country's populace. Science policies in the country have always laid great stress on "Spreading scientific awareness", "Popularizing science and technology" and "Building scientific temper" as their major objectives.

Taking a cue from the country's Constitution and objectives of the Science and Technology Policy Statements, several Non-Governmental Organizations (NGOs) in the country have taken up the challenge of popularizing science in various corners of the country. The government o f India also set up institutes to exclusively promote science popularization and science communication. These include the National Council of Science and Technology Communication (NCSTC) and Vigyan Prasar under the Department of Science and Technology; the National Council of Science Museums, a coordinating body of 27 science museums and science centers across the country; the National Institute of Science Communication and Information Resources (NISCAIR) under the

Council of Scientific and Industrial Research (CSIR), and councils of science and technology in various states for the express purpose of widespread public engagement with science.

An effort has been made here to present the salient features of the four Science and Technology Policy Statements formulated during different time periods in India, with special focus on objectives pertaining to science popularization and how inputs and directions from the Science and Technology Policy Statements over the years have led to establishment of government institutions specifically devoted to science communication, science popularization, and inculcation of scientific temper.

Keywords: Science, Technology, Innovation, S&T Policy, Science Popularisation, Scientific Temper.

1. INTRODUCTION

Science, whether it is being pursued in the laboratories or whether it pertains to the scientific phenomena around us, needs to be popularised, propagated and disseminated among the societies. The reaction of societies to scientific and technological developments and the manner and pace at which new scientific developments are accepted and adopted often defines the pace and direction of human progress.

However, popularising science and spreading scientific temper in culturally diverse societies like India is a complex undertaking. This often involves creating intricate linkages between the practitioners of science and the heterogeneous mass of population. Communication and popularisation of science among the masses in culturally diverse societies therefore always presents a challenge.

It is equally important therefore that attempts and efforts at communicating and popularizing science be backed up by science, technology and innovation policies, plans and strategies. It is imperative that inculcating scientific temper in the society and working towards an informed citizenry be integrated into policy initiatives and also be tempered by the requirements and perceptions of the masses.

India is a highly culturally diverse, vibrant democracy and among the fastest growing economies in the world. Ever since the country's independence, development of science and technology has gone hand in hand with science communication and popularization and inculcation of scientific temper among the country's heterogeneous populace so that it is able to navigate through the complexities that new S&T developments throw up and also contribute to the country's democratic process in a well-informed manner. And, therefore, science policy in the country has always laid great stress on science popularization. In turn, inputs and directions from the Science and Technology Policy Statements over the years have led to establishment of government institutions specifically devoted to science communication, science popularization, and inculcation of scientific temper.

2. INDIA'S TRYST WITH SCIENCE POLICIES

After independence, great stress was laid in India on promoting and institutionalising science and inculcating scientific temper among the country's citizens. Bhikhu Parekh observes, scientific temper, along with concerns like industrialisation, non-alignment, secularism, and national unity became the cornerstone of the national ethos of the independent young nation in 1947 (Bhikhu, 1991).

Following the philosophy whereby dams and scientific institutions were considered the temples of modern India, the Indian state demonstrated a strong pro-science bias resulting in generous state funding for scientific research institutes. Science and technology was sought to be the backbone of a newly independent India. In times to come, national policies enunciated by the Indian state would give science a unique place in the cultural ethos of the nation and the national development programmes.

This was first reflected in the "Scientific Policy Resolution" (SPR) in 1958, passed as an act of Parliament. The policy stated that, "The key to national prosperity, apart from the spirit of the people, lies in the modern age, in the effective combination of three factors, technology, raw materials and capital, of which the first is perhaps the most important, since the creation and adoption of new scientific techniques can, in fact, make up for a deficiency in natural resources, and reduce the demands on capital. But technology can only grow out of the study of science and its applications".

Even though the Scientific Policy Resolution of 1958 did not directly refer to public awareness of science and technology, it however centerstaged science and technology in public policy. One of the concerns of the SPR was to ensure that science is not only practiced by the scientific community, but its impact should be understood by the society. This was expressed under the rubric of 'science, society and development'. Science popularization was seen as a tool for bringing about social reform and making the society mature enough to accept the modern science and technology (Venkateswaran, 2011).

The Constitution of India perhaps holds a unique place among all the constitutions of the world. It is the only constitution that talks of 'scientific temper' laying great stress on the spread of scientific knowledge. The Indian Constitution was amended by the 42nd Constitutional Amendment Act, 1976 by which an Article 51A was inserted making provisions for the Fundamental Duties prescribed for citizens of India. Part IV A, 51A, sub-clause (h) calls on all the citizens of India 'to develop the scientific temper, humanism and the spirit of inquiry and reform' (Varkey, 2000).

Next, in 1983, the Government of India promulgated the "Technology Policy Statement". The document recognised the role and importance of technology in the economic growth of the country and emphasised self-reliance and strengthening of the technology base. The policy document also laid stress on the importance of technology communication to the public. It proclaimed:"… and also of those concerned, at all levels, with any sector of economic, scientific, or technological activity, and not least, the understanding and involvement of the entire Indian people. We look particularly to young people to bring a scientific attitude of mind to bear on all our problems" (Singh, 2014).

With the beginning of a new millennium in 2000, it was time to amalgamate the Scientific Policy Resolution and the Technology Policy Statement. In 2003 was formulated the "Science and Technology Policy" (STP) by the government of India, which laid great stress on spreading scientific temper and popularizing science in the country through effective science communication. The policy states that, "In the half century since Independence, India has been committed to the task of promoting the spread of science. The key role of technology as an important element of national development is also well recognized." Stating that, "People must be able to consider the implications of emerging science and technology options in areas that impinge directly upon their lives, including the ethical and moral, legal, social and economic aspects", STP 2003 outlined the mandate thus: "To ensure that the message of science reaches every citizen of India, man and woman, young and old, so that we advance scientific temper, emerge as a progressive and enlightened society, and make it possible for all our people to participate fully in the development of science and technology and its application for human welfare. Indeed, science and technology will be fully integrated with all spheres of national activity."

STP 2003 further said that "every effort will be made to convey to the young the excitement in scientific and technological advances and to in still scientific temper in the population at large. Special support will be provided for programmes that seek to popularize and promote science and technology in all parts of the country. Programmes will also be developed to promote learning and dissemination of science through the various national languages, to enable effective science communication at all levels."

With technological innovation and invention increasingly being considered important for deriving solutions to some of humanity's intractable problems and for furthering progress, India declared 2010–2020 as the "Decade of Innovation". Innovation was considered significantly important to fuel entrepreneurship and develop new enterprises and hence needed to be woven with science and technology. And thus was born the Science, Technology and Innovation (STI) Policy in 2013.

The STI also recognized the importance of engaging with people as the previous S&T policy documents had done (Singh, 2014). The new paradigm of the Indian STI enterprise is 'Science technology and innovation for the people' under the paragraph 'Capturing Aspirations' which emphasises the importance of the bond between science and society. The STIP identified "Promoting the spread of scientific temper amongst all sections of the society" as a key element. It further gave a call for "triggering changes in the mind-set and value systems to recognise, respect, and reward performances which create wealth from S&T derived knowledge".

The policy underscored the importance of public awareness of science by devoting a whole paragraph to communication of science and technology: "Public understanding of science is an important dimension for introducing and reaching the benefits of modern science and technology to the people. The civilizational aspect of science, or scientific temper, needs to be promoted across all sections of the society systematically. Effective science communication methods by using tools such as National Knowledge Network will be initiated. Public and political understanding of science should be based on evidence and debates with open mind. People and decision makers must be made aware of the implications of emerging technologies, including their ethical, social and economic dimensions".

More recently, once again emphasizing the need for greater public engagement with science, it was for the first time that the annual *Economic Survey 2017–2018* by the NITI Aayog (India's premier planning and policy making organization) incorporated a chapter on Science and Technology (Chapter 8: Transforming Science and Technology in India) with a section on *Greater public engagement of the science and research establishment,* where it emphasized public engagement with science to garner greater support from society.

3. SCIENCE POLICY FOR SCIENCE POPULARIZATION

Apart from nurturing S&T and scientific expertise through establishment of scientific institutions and institutions of higher learning and training scientists, instilling scientific temper among the country's citizens found repeated emphasis in India's Constitution as well as science and technology policy statements over the years. The intent as demonstrated in the policy statements eventually led the government of India to establish institutes and organizations to exclusively promote science popularization and science communication.

These include the National Council of Science and Technology Communication (NCSTC) and Vigyan Prasar under the Department of Science and Technology; the National Council of Science Museums, a coordinating body of 27 science museums and science centres across the country; the National Institute of Science Communication and Information Resources (NISCAIR) under the Council of Scientific and Industrial Research (CSIR), and councils of science and technology in various states for the express purpose of widespread public engagement with science, apart from government media establishments like the All India Radio and Doordarshan.

However, outlined here are three institutions that owe their origin to the strong belief and commitment of Indian policy planners to engage people with science in order to spread scientific awareness and inculcate scientific temper among the country's citizens.

4. NATIONAL COUNCIL FOR SCIENCE AND TECHNOLOGY COMMUNICATION (NCSTC)

Premised on the argument that "Science is both an outlook and a value system" and "the task of creating a scientific temper is a vital necessity for the growth of science and its utilisation in the development process", the sixth five-year plan (1980–85) document tasked that "there has to be dissemination of knowledge about natural phenomena and technological innovations through popular science journals and other media". The plan document also sought to promote "public debate on major issues of science and technology" and "eradication of irrational attitudes, which tend to hold back the nation from the path of progress". Subsequently in May 1982, a new organizational structure, the National Council for Science and Technology Communication (NCSTC), was created to initiate, promote, develop and strengthen public communication of science and technology (Venkateswaran, 2011).

Established under the Department of Science and Technology, Government of India, the plan document detailing the role of the Council stated that NCSTC "should encourage and facilitate all activities aimed at dissemination of science and creation of a scientific temper; more particularly, to orchestrate the various on-going activities in this area so as to render them more effective and mutually supportive." Further, observing that "there are a number of scientifically-oriented voluntary agencies working in rural India", NCSTC was mandated to "provide support to the large number of voluntary science movements and science clubs which have emerged, and would continue to emerge all over the country" (NCSTC, n.d.; NCSTC, DST).

The programmes of NCSTC aim at building capacity for informed decision making in the community and promote scientific thinking. It is devoted towards societal upliftment through the dissemination of scientific knowledge in an informed manner and build programmes with the help of different media which percolate down to every nook and corner of the society.

The NCSTC focuses on outreach activities, training in Science and Technology communication, development, production and dissemination of S&T software, incentive programmes, field based Sci-Com projects, research in S&T communication, international co-operation, motivating students and teachers, environment awareness and programmes with a special component exclusively for women.

Some of its important successful initiatives over the years include the campaigns over the Year of Scientific Awareness, Year of Physics, Year of Astronomy, Year of Mathematics, observation of the National Science Day and National Mathematics Day, the National Children's Science Congress, National Teacher's Science Congress, Science Express etc.

5. VIGYAN PRASAR (VP)

Vigyan Prasar (VP) is an autonomous organization under the Department of Science and Technology (DST), Government of India. Its principal objective is to popularize science through knowledge products in many Indian languages, including publications, films, radio programmes, compendia, portals and a variety of new media, exhibitions and kits (Vigyan Prasar, DST).

Its most recent forays include the *India Science News Feature Service*, a syndicated service featuring Indian S&T developments; the *India Science, Technology and Innovation Portal*, with special emphasis on Indian science, and *DD Science*, a one-hour slot on the national television network, slated to grow into India's first 24-hour science channel.

6. CSIR—NATIONAL INSTITUTE OF SCIENCE COMMUNICATION AND INFORMATION RESOURCES (CSIR-NISCAIR)

When the Council of Scientific and Industrial Research (CSIR) was first established in 1942, one of the basic mandates of CSIR related to the collection and dissemination of scientific information. It was realized very early on that the rapid pace of scientific development and explosion of scientific information could create divisions—among those who know science and those who find it difficult to comprehend the complexities of science. Therefore, great emphasis was laid on communication of scientific information to bridge these gaps and to inculcate scientific temper among the country's citizens.

It is this foresight and vision that led to the establishment of an organization in 1951—the Publications and Information Directorate (PID)—which has today grown to become the National Institute of Science Communication and Information Resources (NISCAIR)—one of the 38 laboratories of the Council of Scientific and Industrial Research (CSIR), India.

While CSIR-NISCAIR is engaged in dissemination of scientific information through research periodicals, monographs, academic publications and encyclopedias, it has also been popularizing science and disseminating scientific information to the masses through books, films and popular science magazines (NISCAIR, CSIR).

NISCAIR has brought out a large number of well-illustrated and moderately-priced popular science books on topics ranging from cells and genes to computers and artificial intelligence, from atoms and composite materials to space technology and stars, and from climate science and solar energy to sports science.

The popular science magazines, however, are the most recognizable faces of CSIR-NISCAIR. The Institute brings out three popular science magazines—*Vigyan Pragati* (Hindi monthly launched in 1952), *Science Reporter* (English monthly launched in 1964), and *Science-ki-Duniya* (Urdu quarterly launched in 1975). These publications are today among the most widely circulating popular science magazines in the country.

Apart from dissemination of scientific information to a wide section of the civil society at a cost they can easily afford, the popular science magazines being brought out by CSIR-NISCAIR also have as aim to bring to light Indian contributions and advances in science, inculcating a rational outlook in the country's citizens, and ensuring wider engagement of the civil society with scientific issues.

7. CONCLUSION

Science policy documents signal intent and commitment to major national goals. India has formulated four science and technology policy statements after independence—Scientific Policy Resolution (1958), Technology Policy Statement (1983), Science and Technology Policy Statement (2003) and the Science, Technology and Innovation Policy (2013). Even a cursory look at the policy statements shows that science policies in the country have always laid great stress on facilitating greater engagement of the civil society with science and technology as revealed in their objectives—"spreading scientific awareness", "promoting the spread of science", "popularizing science and technology" and "building scientific temper".

Besides, the Constitution of India is perhaps the only constitution in the world that talks of 'scientific temper' laying great stress on the spread of scientific knowledge. The Indian Constitution was amended by the 42nd Constitutional Amendment Act, 1976 by which an Article 51A was inserted making provisions for the Fundamental Duties prescribed for citizens of India. Part IV A, 51.A, sub-clause (h) calls on all the citizens of India 'to develop the scientific temper, humanism and the spirit of inquiry and reform'.

Science popularization and effective communication of scientific information can play a significant role in strengthening the functioning democracy that India is by ushering in scientific literacy. High scientific literacy rate is important for today's technologically-driven society. As Christie Wilcox says: "When only a small percentage of our populace—including our policy-makers—has a firm grasp on the science behind the debates, we are doomed to make grievous errors in our decisions on a wide variety of issues, from climate change and genetically modified foods to stem cell research and public health and vaccinations" (Wilcox, 2012).

Therefore, with science becoming increasingly more complex and all pervasive, science policy statements need to factor in intentions and directions to popularise and propagate scientific developments among the masses. Some initiatives pertaining to popularising science and strengthening science communication are suggested as probable policy inputs for future science policy statements:

1. Scientists may be encouraged to undertake public outreach.
2. Create mechanism for sustained interactions between scientists, science communicators, journalists and policy makers/authorities.
3. Encourage greater engagement of the scientific community with the public by incentivising public engagement in the form of career progression, etc.
4. Integrate public outreach/engagement in publicly funded scientific projects.
5. Formulate national-level surveys to gauge/track/measure understanding of scientific issues among the public for designing

appropriate science communication activities and programmes for targeted groups.

6. Government science institutions/organisations should strive for professional media outreach by employing trained science communicators.
7. Media channels should be sensitised and motivated to enhance science content.
8. Science media literacy may be incorporated into school curriculum.

The relevance and significance of science communication can never be obsolete. With increasing scientific complexities and technological developments, the need for demystification and communication and propagation of scientific ideas to the civil societies will always remain relevant.

ACKNOWLEDGEMENT

I would like to thank Director-General, Council of Scientific and Industrial Research (CSIR), India for nominating me for the conference.

REFERENCES

[1] Bhikhu, P. (1991). Nehru and the National Philosophy of India. *Economic and Political Weekly*, 26(1/2), 35–48.

[2] NCSTC (n.d.). *National Council for Science and Technology Communication NCSTC.* From http://www.dst.gov.in/scientific-programmes/st-and-socio-economic-development/national-council-science-technology-communication-ncstc

[3] Singh, B.P. (2014). Science Communication in India: Policy Framework. *Journal of Scientific Temper, 2*(1&2), 141–151.

[4] Varkey, M.J. (2000). Science popularization and the Indian constitution-Correspondence. *Indian Academy of Sciences*, 1054–1054.

[5] Venkateswaran, T.V. (2011). Infrastructure for public communication of science in India. *Constructing Culture of Science*, 37–65.

[6] Wilcox, C. (2012). It's Time to e-Volve: Taking Responsibility for Science Communication in a Digital Age. *Biol. Bull., 222*, 85–87.

Policy Making on the Popularisation of Science, Technology and Innovation (STI) in Nigeria

S.L. Wali
Department of Policy Division, Raw Materials Research and Development Council, Abuja, Nigeria
E-mail: ceo@rmrdc.gov.ng

ABSTRACT: Over the years, the Nigerian Government has initiated some activities to promote, develop and market the STI policy in order to particularly enhance competitiveness in raw materials and products development as championed by the Raw Materials Research and Development Council (RMRDC) situated in the central part of the Federal Capital Territory (FCT), Abuja.

Some of these activities include: Organization of Science quiz; Provision of Technical education projects in Schools; Organizing Technology Expositions, Organizing TV Program on Science discoveries; Establishing Resource Centres; Nigerian Raw Materials Exposition (NIRAM EXPO); International/Local Co-operation and Linkages; Collaboration with Public and Private Sector on Industrial Development; Promoting participation of Women in Science and Technology; among others etc.

Keywords: Policy, Science, Technology, Innovation, RMRDC.

1. INTRODUCTION

1.1 Nigeria at a Glance, *see Figure 1.*

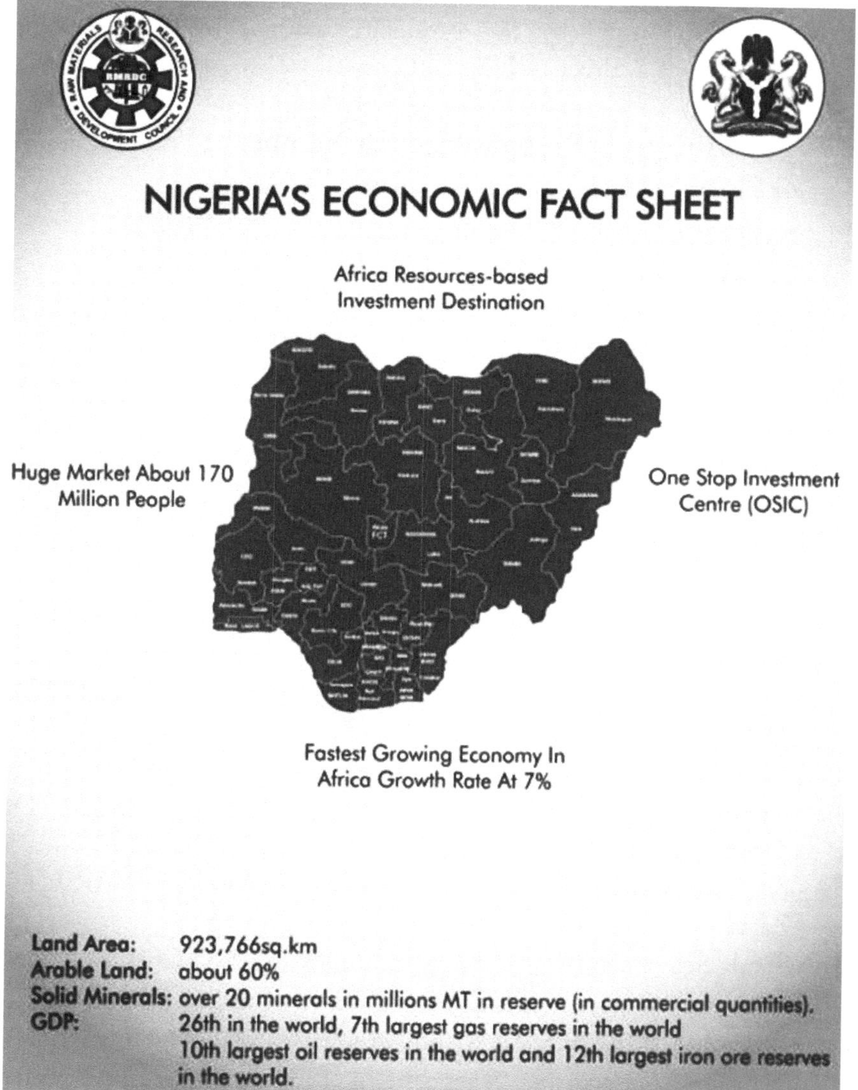

Figure 1: Nigeria's Economic Fact Sheet

2. ESTABLISHMENT OF FEDERAL MINISTRY OF SCIENCE AND TECHNOLOGY (FMS&T)

The Federal Ministry of Science and Technology is a Nigerian ministry and was established to facilitate the development and deployment of science and technology apparatus to enhance the pace of socio-economic development of the country through appropriate techno-logical inputs into productive activities in the nation. It is headed by a Minister appointed by the President, assisted by a Permanent Secretary, who is a career civil servant.

2.1 The Ministry is Responsible for a Number of Parastatals, or Government-Owned Agencies

1. National Board for Technology Incubation (NBTI)
2. Energy Commission of Nigeria (ECN)
3. The Nigerian Institute of Science Laboratory Technology (NISLT) - www.nislt.gov.ng
4. Nigerian Institute for Trypanosomiasis and Onchocerciasis (NITR)
5. National Biotechnology Development Agency (NABDA)
6. National Centre for Technology Management (NACETEM)
7. National Office for Technology Acquisition and Promotion (NOTAP)
8. Nigerian Natural Medicine Development Agency (NNMDA)
9. National Space Research and Development Agency (NARSDA)
10. Raw Materials Research and Development Council (RMRDC), Abuja
11. Nigerian Building and Road Research Institute (NBBRI)
12. National Institute of Leather Science and Technology (NILEST), Samaru Zaria
13. National Research Institute for Chemical Technology (NARICT), Zaria
14. Sheda Science and Technology Complex (SHESTCO), Abuja
15. Project Development Institute (PRODA), Enugu
16. Federal Institute of Food & Industrial Research, Oshodi (FIIRO)
17. National Agency for Science and Engineering Infrastructure (NASENI), Abuja.

2.2 About RMRDC

The Raw Materials Research and Development Council (RMRDC) is an agency of the Federal Government of Nigeria vested with the mandate to promote the development and utilization of Nigeria's industrial raw materials. The decision to set up an agency for raw materials development originated from the recommendations of a workshop on industrial matters organized by the Federal Ministry of Industry, Manufacturers Association of Nigeria (MAN) and the Nigerian Institute of Social and Economic Research (NISER) in July 1983. The Council was established by Act 39 of 1987, but commenced operation on February 10, 1988. Nigeria's focal point (Figure 2) for the development and utilization of the nation's vast raw materials.

Figure 2: RMRDC Building

2.2.1 *Mandates of RMRDC*

1. To draw up policy guidelines and action programs on raw materials acquisition, exploitation, and development.
2. To review from time to time raw materials resource availability, and utilization to advise of the federal government on the strategic implication of depletion conservation or stock-piling of such resources.
3. To advise on the adaptation of machinery and process for raw materials utilization.
4. To encourage publicity of research findings and other information relevant to local sourcing of industries;
5. To encourage the growth of implant research and development capabilities;

6. To advise on and devise award systems for industries that achieve any breakthrough or makes innovations and inventions;
7. To organize workshops, symposia and, seminars from time to time designed to enlighten people on new development and solutions discovered;
8. To consider and advise on special research grants for specific objectives; and
9. To consider and advise on any other issue capable of enhancing the objectives of the Council.

2.2.2 *Vision of RMRDC*

To be an indispensable catalyst for industrial growth and development in Nigeria.

2.2.3 *Mission of RMRDC*

To Promote the Development and Optimal Utilization of Nigeria's Raw Materials for Sustainable Industrial Growth.

3. STI (SCIENCE, TECHNOLOGY AND INNOVATION) POLICY IN NIGERIA

In 2005, the need to carry out a system-wide reform was consummated and implemented under the Nigeria/UNESCO Science, Technology and Innovation (STI) reform initiative. It adopted the National Innovation System (NIS) approach as a framework for Science, Technology and Innovation (STI) system reform. The reform, among others issues, stressed that economic development initiatives, institutional governance, research and development (R&D) agenda for the country, funding mechanisms, Intellectual Property (IP) and STI Infrastructure development be addressed in any revised STI policy.

One notable feature of this policy is the emphasis on 'Innovation', which has become a global tool for fast-tracking sustainable development. This policy is a clear demonstration of the country's renewed commitment to ensure that our R&D engagements enhance new business development, encourage employment generation, wealth creation through the creation and growth of SMEs that are ultimately translated into goods and services in the market place.

3.1 Policymaking on the Popularization of STI in Nigeria

This Draft National Science, Technology and Innovation Policy thus have as its core mission the evolution of a new Nigeria that harnesses, develops and utilizes STI to build a large, strong, diversified, sustainable and competitive economy that guarantees a high standard of living and quality of life to its citizens. Specifically, the STI policy was designed to provide a strong platform for science, technology and innovation engagement with the economic transformation that is citizen-centered. The concept of innovation is strongly welded into this new framework of actions, policies, and programs to apply science and technology to achieve social and economic objectives. This has come from the realization worldwide that technologies can bring about desired changes only when they are fully integrated into local systems and practices which may not necessarily have emerged from R&D. In this sense, innovation is a critical driver for socio-economic and sustainable development.

To effectively foster a seamless engagement of STI with the desired transformation, there is a need to popularize STI policy across all sectors of the Nigerian economy. This will ensure the integration of science, technology and Innovation into the nation's national development plan thereby making STI a driver to the achievement of sectoral goals, objectives and programs.

The popularization of Science, Technology and Innovation should be considered as an essential part of the overall Science, Technology and Innovation (STI) policy with its strategic objectives and an action plan to act as a guide in the implementation of the set objectives, this will maintain and strengthen support by the general public for science, technology, and innovation.

3.1.1 *Objectives of the Policy for the Popularization of STI*

- To provide information about science and technology in general, and about scientific-technological research in particular.
- To raise awareness of the importance of scientific and (innovative) technological research and innovation.
- To account for the use of the budget that the government allocates to such research and innovation.
- To create a culture that welcomes technological innovation.

4. ACTIVITIES OF THE RMRDC IN THE POPULARIZATION OF STI POLICY IN NIGERIA

4.1 Organization of Science Quiz

In a quest to sustain awareness on science innovations and its creative exploitations in the entrepreneurial abilities in youngsters at the secondary schools, an annual quiz competition has been conceptualized under the Catch Them Young Programme setup by the RMRDC (Figure 3(a), (b) & (c)).

Figure 3(a), (b) & (c): Highlights of the Science Quiz

4.1.1 *Advantages of the Quiz Competition*

The objectives of the competition are as follows:

1. To create awareness on the importance of raw materials and the role of RMRDC as a government agency with mandates for the development of the nation's raw materials resources for use by industries.
2. To inculcate the spirit of research in raw materials in young school children.
3. To carry out competitive quiz contests amongst schools on raw materials-related topics.
4. To increase the creative entrepreneurship ability of students.

4.2 Provision of Technical Education Projects in Primary and Secondary Schools

The ultimate aim of the project is to provide existing and would-be teachers and the whole school environment with the skills, equipment, and tools they require to make them as comfortable as possible with technological education and thus teach it better. The project is directed at the intermediary level (i.e. the actors involved, both in the field of education as in the field of the popularisation of science, technology and innovation) and will endeavour to provide a framework for technological education and provide the necessary, essential building blocks for working out a curriculum for the subject. By involving the intermediary level some of whom are responsible for implementing the curriculum in the classroom greater support will be elicited in the field, raising the prospect of more effective implementation. The framework established in this way will also be used to screen and validate new projects within the action plan, applying the principles behind technological education to a broader area of and using them as leverage to enhance the plan's structure, to improve the coordination of the activities within it, and to promote more cooperation between the actors.

4.3 Organizing Technology Expositions

The aim here is to showcase expertise in new technologies to the general public and to highlight the full range of activities involved in technological innovation, ranging from conceptualization to the design and realization of new products or services bringing innovation and its benefits to the attention of the general public, see Figures 4(a) & (b).

Figure 4(a): Glance from the Recent Technology Exposition

Figure 4(b): Honourable Minister of Science and Technology, Hon Ogbonnaya Onu, with the DG. RMRDC, Prof. H.D. Ibrahim

This is also a flagship programme to provide the opportunity for synergy among stakeholders in the Science, Technology, and Innovation sector was by inventors, investors and government MDA's collaborate to STI in Nigeria.

4.4 Organizing TV Programmes on Science Discoveries

Broadcast on both Radio and Television, featuring exciting documentaries about recent scientific achievements that are closely related to the everyday experiences of most citizens. Each edition of the program will be built around a prominent scientist behind an important scientific breakthrough in their specialist field. It will also involve a science program in which children aged 10 through 12 try to answer a scientific or technological question, competitions will also be held to give youngsters from 5th and 6th years of primary school the chance to find their own answers to a question that particularly interested them.

Also, a series of short, 5-minute films explaining examples of technological innovation to the general public in readily comprehensible language.

4.5 Establishing Resource Centres

The Raw Materials Resource Centre (RMRC), which is already operational in the Council's headquarters in Abuja, a multimedia-based, combining both the physical, electronic and pictorial displays of raw materials and provide in addition, data on all matters, relating to raw materials sourcing, development and utilization. It is also designed to provide a critical linkage between the Research Institutions (RIs) and end-users of R&D information, through data sharing via the Raw Materials Information System (RMIS). The resource centre, therefore, has in addition to the traditional RMDC, the RMIS, the electronic/ pictorial display and VSAT facilities providing the expected linkage among all the RIs in the country, hence the change in nomenclature to Raw Materials Resource Centre. Features of the new concept include a substitution of wooden display cabinets and billboards with aluminum display cabinets and expandable stand. Introduction of roll-up banners, replacement of perishable items with scale models, documentaries, etc.

The Resource centers that offer delightful, exciting and richly innovative discovery experience for enthusiasts of all ages through interactive exhibits to help visitors rediscover the science and technology behind familiar objects and inventions, see Figures 5(a), (b) & (c).

(a)

(b)

(c)

Figures 5(a), (b) & (c): RMRDC Display Centre Commissioned by the Honourable Minister of Science and Technology, Hon Ogbonnaya Onu, with the DG. RMRDC, Prof. H.D. Ibrahim

4.6 Nigerian Raw Materials Exposition (NIRAM EXPO)

NIRAM Expo (Figure 6(a) & (b)) is an event designed to focus national attention on critical challenges facing the application of Science, Technology and Innovation (STI) in manufacturing industries, which borders on securing a sustainable approach in sourcing adequate resources and creating efficient linkages/synergy among the industry players in the economy.

It is also programme to provide an opportunity for synergy among stakeholders in the Science, Technology and Innovation sector were inventors, investors and government MDA's collaborate to STI in Nigeria.

Figure 6(a) & (b): The Representative of the Vice-President, Dr. Olajumoke Oduwole & Dr. Frank Udemba Jacobs President of MAN Round the RMRDC Exhibition

4.7 International/Local Co-operation and Linkages

The international scene is linked in such a way that exchanges, partnerships, networking and co-operation have become the norm for keeping abreast with the pace of innovation in all sectors of the economy. Given the scope for international co-operation (Figure 7(a) & (b)) in science and technology at the regional and sub-regional levels, the Government will adopt all necessary measures to consolidate and widen bilateral and multilateral programmes to the benefit of its science and technology plans, especially the flow of scientific and technological information.

(a)

(b)

Figure 7(a) & (b): The DG. RMRDC, Prof. H.D. Ibrahim and Some International Collaborators

4.8 Collaboration with Public and Private Sector on Industrial Development

Some collaborative efforts of the Council with the private and public sectors of the Nigerian economy include:

4.8.1 *Production of Calcined Kaolin as Partial Substitute for Titanium Dioxide in Paint Making*

Titanium dioxide (TiO_2) is one of the pigments used in the paint industry. This compound is 100% imported into the country with over 26 million USD spent annually in foreign exchange.

Given the above, the Council embarked on Research and Development to develop a substitute that will reduce/minimize foreign exchange expenditure on this compound, at least up to 40% reduction level. Apart from its use in paint formulation as a partial substitute to titanium dioxide, calcined kaolin is also used as an extender in the manufacture of cement. Therefore, its utilization in cement production will create employment opportunities in its value chain from mining, beneficiation, and processing.

Based on the success recorded at the laboratory level and the successive industrial testing, the Council commissioned a local engineering company (Bon Affaire Company Nigeria Ltd.) to fabricate a 10 Ton/day calciner (Figures 8(a), (b) & (c)). The fabrication of all parts

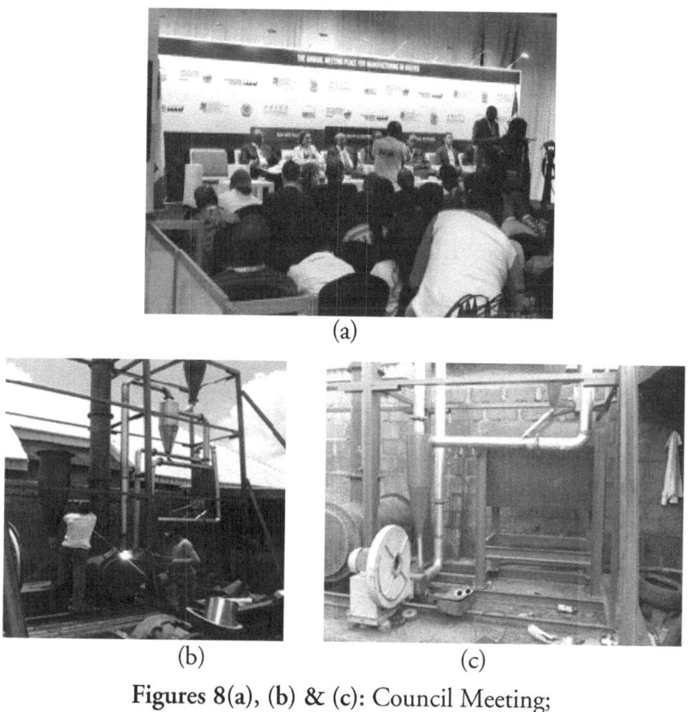

(a) (b) (c)

Figures 8(a), (b) & (c): Council Meeting; Gas suspended Calciner under Construction has been completed and awaiting deployment to the RMRDC demonstration centre, The Council has called on the Paint Subsector to commercialize the project after completion to boost the supply of calcined kaolin as partial substitute to TiO_2 in Nigeria.

Dangote Cement Company has indicated interest in the utilization of the calcined kaolin as an extender in cement production.

4.8.2 *Development of Water-Borne Paint from Blends of Natural Rubber Latex and Poly Vinyl Acetate*

The polyvinyl acetate used in the production of paints in Nigeria is imported from China, India, Singapore, etc. In 2010, Nigeria spent $1.22 m on the importation of polyvinyl acetate and this accounts for 0.75% of global importation of this compound, making Nigeria the third largest importer of polyvinyl acetate in Africa. In 2009, RMRDC commissioned the project to reduce the level of importation of the material into the country for paint production. Optimum blend formulation was developed which was used to produce emulsion paints,tested and found to compare favurably with the commercial paints available in the market.

The project has created a new area in the utilization of natural rubber latex in the country. The sustainability of the supply of the natural rubber latex for paint production is guaranteed because there are rubber plantations, rubber processing centres and farmers in the eleven rubber producing states in the country.

The project which was carried out in collaboration with the Paint Manufacturers Association of Nigeria has been patented and is ready for commercialization for which the Paint Manufacturers Association has indicated interest.

4.8.3 *Development of Pilot Plant for the Production of Caustic Soda and Precipitated Calcium Carbonate*

Sodium hydroxide and precipitated calcium carbonate (PCC) which are versatile industrial chemicals are 100% imported. As of 2010, 98,869,774 kg of sodium hydroxide was imported and this amounted to 7 million USD. Equally, 22,985,080 kg of PCC valued at around 4 million USD (NBS 2010) was imported for the same year.

Given the availability of the raw materials for the production of this chemical, the Council embarked on the precipitation method for co-production which presents great economic potential as both products are valuable with a wide range of applications.

The major raw material is hydrated lime which is produced from limestone and there are large deposits (over 180 million tons) of this mineral in the country. Research into their potentials as raw materials for the production of calcium oxide has been investigated. Sodium carbonate is another major raw material used in the production of PCC (Figure 9).

Figure 9: Caustic Soda and PCC Pilot Plant

4.8.4 *Development of Water-Based Thermostable Vaccine for the Control of Newcastle Disease in Rural Poultry*

The Nigerian poultry population has been estimated at 150 million and in terms of revenue, it has been ranked second only to the oil industry. The poultry production system in Nigeria comprised both commercial and rural scavenging poultry. The rural scavenging poultry accounts for about 80% of the total poultry population with little or no input in the production system from both the owners and the government. Rural poultry farming is a very important means of alleviating poverty, but production has always been hampered by frequent outbreaks of Newcastle diseases, arising from ineffective delivery of vaccines due to the peculiar local environment.

A thermostable vaccine that can withstand extreme temperatures of up to 45°C for about one week has been produced in collaboration with the National Veterinary Research Institute (NVRI), Vom and the product is now available in lyophilized form at NVRI. RMRDC and NVRI are currently working out the modalities for carrying out popularization of the thermostable vaccine among rural farmers in the South-western Nigeria and subsequently other geopolitical zones of the country. This is to boost poultry production and increase income to rural farmers.

4.8.5 *Development/Beneficiation of Gypsum for POP and Cement Manufacturing*

Gypsum which is utilized in the cement and POP production is 75% imported by the Cement industry and 25% sourced locally. One of the reasons for the low local utilization is the attendant high impurity associated with the mined gypsum. Based on this, the Council embarked on its beneficiation to meet industrial specifications for POP and Cement production.

The import of various grades of gypsum and POP in 2008 was 6.7 million USD. Between 2011–2013, import figures for gypsum, POP and other associated products rose to about 9.3 million USD. The national demand for gypsum for cement production alone is over 1 million MT/Annum.

Figure 10: Gypsum Scrubber Developed and Fabricated by RMRDC

This project is aimed at developing local engineering capacity for the design and fabrication of POP process equipment (Figure 10) to promote the development of POP production in Nigeria. Also, it is to ensure the import substitution of POP by 15–20%.

The Council fabricated a Scrubber for washing of gypsum to get rid of impurities from the mined gypsum. Physico-chemical analysis was also conducted to ascertain its purity level and composition for POP and Cement production. This scrubber has the capacity to wash 30 tonnes of gypsum per day. It has been demonstrated, commissioned and handed over to Yobe State Gypsum Miners Association.

4.8.6 *Promotion of Castor Oil and its Derivatives*

Castor is a commercially important strategic plant because of the many industrial applications of the oil obtained from the seed and also the seed cake which is a source of organic fertilizer. Castor oil finds application in almost all the sectors of the manufacturing industries. From the automobile industry where it is used in hydraulic and brake fluid production (Figure 11(a) & (b)) to the pharmaceutical where it is used to produce laxative. The highest application is found in the cosmetic sector in the formulation of various body and hair treatments and soap production.

Figure 11(a): Replicated Sheller

Figure 11(b): Imported Sheller

4.8.7 *Development and Production of Web Offset Ink in Nigeria*

The web offset ink is one of the most consumed types of printing ink in the country. It is usually in a paste form with fine rheology. Web-offset ink is an industrial raw material that is at the centre stage in news production. News production using newsprint cannot be completed without the web-off set ink.

The development and production of web-offset ink, Figure 12 (using about 85% local raw materials) for the media has been completed and awaiting commercialization which will result in quality web-offset ink that will match the imported brand and the problem of supply gap created by importation will be solved. The economy will be boosted and foreign exchange will be saved and the much needed employment opportunities will be created.

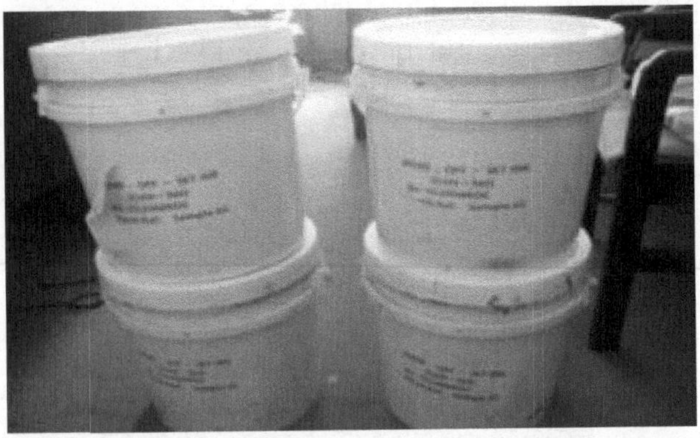

Figure 12: Web Offset Ink Produced from Local Raw Materials

4.8.8 *Development and Technology Domestication of Mini-Refinery for Naphtha Hydro-Treating and Fluid Catalytic Cracking Units.*

This project seeks to domesticate the technology for naphtha hydro-treating and fluid cracking units in Crude oil refining. The specific objective includes:

- Study of various Naphtha Hydro-treating and Fluid Catalytic Cracking processing routes employed in the Nigerian Refineries.
- Process selection and preliminary design of the refinery.
- Detail design of Naphtha Hydro-treating Unit and Fluid Catalytic Cracking.

- Fabrication of the design refinery units.
- Assembly of fabricated units.
- Test-running of the assembled unit.
- Plant commissioning.

Most of the components required to build the cracking unit have been fabricated and assembled. These include surge drum, heat exchanger, condensers, low pressure separator, NHU, reactor, furnace, burners, column, dis-engander, regenerator, cyclone, air grid and fluid catalytic cracking unit riser. The Naphtha Hydro-treating unit (Figure 13) is expected to produce one barrel of refined oil per hour while the fluid catalytic cracking unit is expected to produce 5 barrels of refined oil per batch (Batch time is nine hours).

Figure 13: Completed NH Unit

4.8.9 *Production of School Laboratory Chemicals and Reagents Project*

There are over 500 different industrial chemicals presently being utilized in the country, out of which there are over 70 commonly used secondary school chemicals. It is needless to state that all these chemicals are imported although the country has comparative advantages in terms of availability of the raw materials to produce some of them. A huge amount of money is being expended on the importation of these chemicals for educational purposes in secondary schools and tertiary institutions in Nigeria.

The teaching of science subjects in our secondary schools and colleges has continued to experience a steady decline in performance because most science laboratories lack chemicals and reagents in the right quality and quantity, the result of which usually leads to mass failures in examinations with attendant implication for science education and technology acquisition.

It was in this regard and to motivate the production of laboratory chemicals and reagents in Nigeria that the Council facilitated and successfully completed a pilot plant for the production of Laboratory chemicals, see Figure 12. The plant was commissioned at NARICT, Zaria and has the capacity to produce: Copper Sulphate, Magnesium Sulphate, Zinc Sulphate, Copper Carbonate, Iron Sulphate, Ammonium Sulphate and Calcium Chloride. The patenting of this technology is in progress.

Figure 12: Pilot Multi-Chemical Plant for the Production of Laboratory Chemicals and Reagents

4.8.10 *Dairy Projects*

Nigeria's continued dependence on imported milk powder for processed milk and other dairy products has not only affected the national economy through an estimated foreign exchange expenditure of 10 million USD annually, it has slowed down the development of the indigenous dairy industry. The cattle population in Nigeria is estimated at 19 million, however, 17.5 million (92%) is owned by seminomadic cattle rearers whose herds are characterized by low production and productivity as a result of low genetic potentials of the local breeds, poor quality feeds, disease and inadequate veterinary care. Indigenous milk production, processing and marketing among this group is generally not organized and over 90% of fresh milk produced does not undergo adequate processing. In view of this, the Council embarked on a number of diary projects in order to reduce the nation's dependence on imported milk and to develop the local dairy industry in Nigeria.

The Council in collaboration with the Agriculture and Rural Develop ment Department of the FCT (ARD-FCT), United States Agency for International Development (USAID) – Markets, Lands-o-Lakes Inc. and the National Livestock Project Department, established a Milk Collection and Cooling Centre (Figure 15) to serve some Cooperative Associations at Paikon kore, Gwagwalada in the FCT.

Figure 15: Milk Processing Facility at Gwagwalada, FCT

4.8.11 *Production of Wood Seasoning Kilns for Optimal Drying of Wood to Requisite Moisture Content*

The Nigerian wood industry has been having problems with assessing adequate process equipment since 1980's as a result of the downturn of the economy. The majority of the current operatives in the sawmill and furniture industries are cottage and small scale industries with 3–5 men workforce. They mostly employed crude implements and sub-standard drying procedures. A major objective of this initiative is to build local capacity in wood process equipment fabrication. Another important objective is to standardize wood processing in Nigeria along global best practices.

Under this project, RMRDC collaborated with Palcon Nig. Ltd., a private sector engineering firm to design and fabricate wood seasoning kilns for use by local industries. This was to ensure the domestication of wood seasoning technology and the upgrading of locally produced wood products to a global standard.

The two types of kilns (an all-electric wood seasoning kiln and a mechanically fired wood seasoning kiln) were successfully designed and fabricated as shown in Figure 16(a) & (b). The kilns have been tested and found to be effective in reducing the wood moisture content to as low as 7%.

(a)

(b)

Figure 16(a) & (b): Wood Processing Technology: Wood Seasoning Kiln Developed in Collaboration with Palcon Nigeria Limited, Enugu

4.8.12 *Production of Prototype Small Scale Automated Weaving Loom*

Traditional weaving in Nigeria is laborious and cumbersome. Consequently, a proposal was submitted to the Council by a research team from Ahmadu Bello University Zaria. The objective of the project was to bring the small scale automated weaving loom to the door steps of local weavers to enable them to meet the production targets and produce more competitively Figure 17(a) & (b).

Further to this, an automated weaving loom (Figure 17(c)) was designed, fabricated and test run after modifications. The equipment has been deployed for further evaluation, after which it will be popularized. The patenting of this technology is in progress.

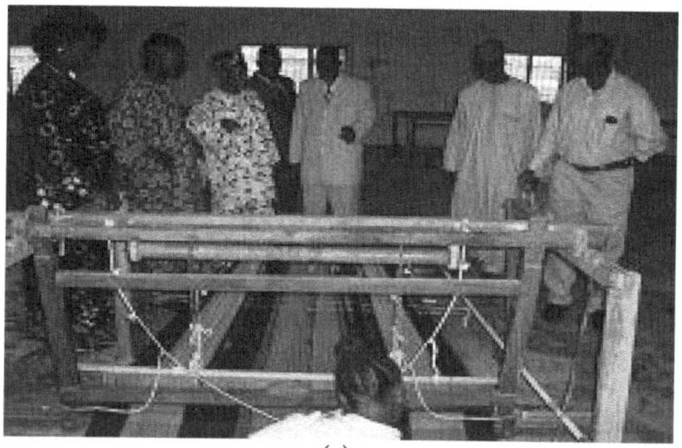

(a)

(b)

Figure 17(a) & (b): Silk Weaving Loom

Figure 17(c): Akwete Weaving Machine, Textile Processing Technology: Developed by RMRDC for Akwete and Silk Weaving Cooperative Societies

4.8.13 *Development of Moringa oleifera for Water Treatment, Nutritional and Medicinal Uses*

In 2008, the Council initiated a sensitization workshop and exhibition on the socio-economic uses of local plant resources such as *Moringa oleifera*. Consequently, the Council in collaboration with the Chemical Engineering Department of Ahmadu Bello University (ABU), Zaria, evolved and embarked on two R&D project lines in the utilization of *Moringa oleifera*; the first being the use of *Moringa Oleifera* seeds as raw materials for Water Treatment with ABU, Zaria, while the second, a collaboration with Nnamdi Azikiwe University, Awka is on exploitation of the nutritional and medicinal values of *Moringa oleifera*.

The outcome of these collaborations has led to the following achievements:

- Development of laboratory model and Pilot plant on Moringa water treatment, Figure 18(a) & (b);
- Two Post Graduate students (M.Sc in Moringa oil extraction and Ph.D in Water Purification using Moringa coagulant);
- Development of products such as Moringa tea, Moringa powder and Moringa oil;
- Two National Summit on *Moringa oliefera* with proceedings produced;
- Formation of Moringa Development Association of Nigeria registered with the Corporate Affairs Commission;
- Equipment support to some Moringa SMEs.

Figure 18(a): Moringa Water Treatment Pilot Plant-Mild Steel Laboratory

Figure 18(b): The 10,000 Litre Water Treatment Pilot Plant

4.8.14 *Upgrading/Optimizing Killishi Processing Technology*

Nigeria consumes 360,000 tons of beef annually. However, it has been predicted that this quantity will rise to 1.3 million by 2050 due to increasing population growth. The 19 million heads of cattle in Nigeria do not meet local demand for beef. Out of the total 360,000 tonnes of beef consumed annually, approximately 20% is used for kilishi production. Kilishi is a well-known sun-dried meat product that undergoes light smoke-drying process. Kilishi has over the years gained popularity among Nigerians as a light snack. However, the traditional process technology involves drudgery, taking 2–3 days to complete a production cycle; products are also of low aesthetic quality and limited shelf life. The processes are unhygienic and vary from one processor to the other or even from batch to batch from the same processor, due to poor scientific and technical standards of the production process. In view of this, the Council saw the need to establish a Kilishi processing plant that is based on an efficient technology which produces wholesome Kilishi product. Consequently, RMRDC in collaboration with the University of Benin developed a Kilishi processing technology and fabricated the equipment. The initial technology is now being upgraded to attract investment in modern Kilishi processing.

This process technology is based on factory type production system (Figure 19) using dryers, smoking kiln, unit control measures of inputs and high sanitary conditions, to cut production time to less than 10 hours. The objective of this project is to promote investment in modern kilishi production.

Figure 19: RMRDC Kilishi Model Factory

The technology has been well accepted and has been used to establish the processing plant by RMRDC. A piece of land has been acquired from the African University of Science and Technology, Abuja to house the project. The factory building has been completed; the equipment has been fabricated installed and test-run.

Trial production was carried out and samples produced were used to conduct a sensory evaluation of the product by the staff of the Council. The plant was relocated to Suleja, Niger State, ready for commissioning and commercial-scale production. This technology has been patented by the Council

4.8.15 *Development of Automotive Brake Pad Using Palm Kernel Shell*

Break pad and the lining is an automobile braking system (friction material) with a demand in Nigeria currently put at 33,049,600 pieces (pcs) per annum and costs the country over 5 million USD on importation. The raw materials for the production of brake pad and linings are readily available locally. Nigeria has companies producing an average of 120,000 pcs yearly, which is about 70% of installed capacity per company. Seven out of the eleven local brake pad and lining manufacturing companies identified in the country have closed down according to survey data.

Though there are few automobile assembly plants in Nigeria, the country remains a net importer of motor vehicles and their spare parts.

This has resulted in huge foreign exchange losses for the country. Also, the conventional asbestos-based brake pad is found to be carcinogenic. In order to reverse this trend, Council in collaboration with Obafemi Awolowo University, Ile-Ife developed Palm kernel shell-based brake pad and lining. The projects aims to increase the utilization of local raw materials and reduce the importation of brake pads.

The project has been completed and samples of brake pads (Figure 20) produced met Standards Organisation of Nigeria and United States standards for brake pads. Currently, the process for patenting is ongoing while field testing of 100 pairs of the product (PKS brake pad) has been carried out with satisfactory results. Public presentation and commercialization is the next step being undertaken on this project.

Figure 20: Samples of PKS Break Pad

4.8.16 *Cashew Nut Shell Liquid (CNSL)*

The Council procured some CNSL machines (Figure 21) in order to utilize the wasted cashew nuts shell for the production of Cashew Nut Shell Liquid (CNSL). The cashew nutshell liquid is used for the following purposes:

- Paints,
- Laminating resins,
- Rubber compounding resins,
- Cashew cements,

- Polyurethane based polymers,
- Surfactants,
- Epoxy resins,
- Friction linings,
- Foundry chemicals.

Figure 20: Cashew Oil Press Machine

The machine is for the extraction of oil from cashew nuts, the Nigerian government through the RMRDC has donated four (3) cashew machines to four (3) different Tertiary Schools such as:

1. Ibrahim Badamasi Babangida University, Niger State
2. Abia State University, Abia State
3. Kogi State University, Kogi State
4. Federal University of Agriculture Abeokuta (FUNAAB), Ogun State.

4.9 RMRDC Publications on Science and Technology Findings

4.9.1 *Policy Briefs*

The Council in its efforts in providing policy direction to the development of key strategic raw materials, has published Policy Briefs on some selected raw materials/products to ascertain their level of development and provide the road map for their exploitation. Policy Briefs on the following raw materials/products had been prepared:

- Abrasive
- Abrasive tape
- Latex
- Bolt and Nut
- Shoe Soles
- Cosmetics
- Conduit Pipe
- Colourant
- Printed Circuit Board
- Herbs
- Citric Acid
- Cashew nut shell oil
- Soya bean oil
- Hinges and Keys
- Seat Foam
- Seat Cover
- Tires and Tubes
- Salt
- Mica
- Polyester
- Tea and Coffee
- Industrial Starch
- Petroleum jelly
- Textile clothing and Garment
- Bricks
- Tissue paper
- Marbles and Granite
- Gum Arabic
- Nails
- Jute sack
- Pulp and paper
- Seasame seed
- Shea butter
- Glucose
- Sulphuric acid
- Hides and skin (Leather)
- Tomatoe puree
- Kenaf

4.9.2 ***Journal of Raw Materials Research*** promotes research and development on local raw materials for industrial utilization. The journal covers the fields of agriculture, minerals, new and advanced materials, petrochemicals and engineering. Greater emphasis shall be laid on applied research findings that have the potential application to the improvement of industrial processes particularly those that relate to:

1. Processing of mineral and agricultural raw materials, new and advanced materials and petrochemicals into economically viable products.
2. Substituting identified major or minor industrial inputs with products of equivalent value.
3. Development of process technology or fabrication of machines and other and other components particularly through the adoption of local technology.
4. Any other research findings that are considered relevant to the objectives of the journal by the editorial board.

4.9.3 ***Raw Materials Update*** a bi-annual publication of the Council. It deals with raw materials sourcing and utilization in order to promote backward integration and the creation of jobs and wealth.

4.10 Promoting Participation of Women in Science and Technology

Traditionally women have not been encouraged to embrace the learning of science at all levels of the educational system. This has created a lopsided proportion of women involved in the learning, teaching and practice of science, engineering, technology and related fields as compared to their male counterparts.

The Council in her resolve to contribute to local contents development, increase foreign exchange earning using Nigeria's local resources, increase GDP, employment generation and ultimately economic diversification has ear-marked shea-butter development as one of the strategic projects to achieve it. Furthermore, to actualize the above resolve, partnership with relevant private sectors and non-governmental organizations (NGOs) is inevitable; hence the collaboration with the National Shea Products Association of Nigeria (NASPAN), see Figure 21(a), (b) & (c).

(a)

(b)

Figure 21(a) & (b): RMRDC Shea Butter Training for Woman Using Modern Technology

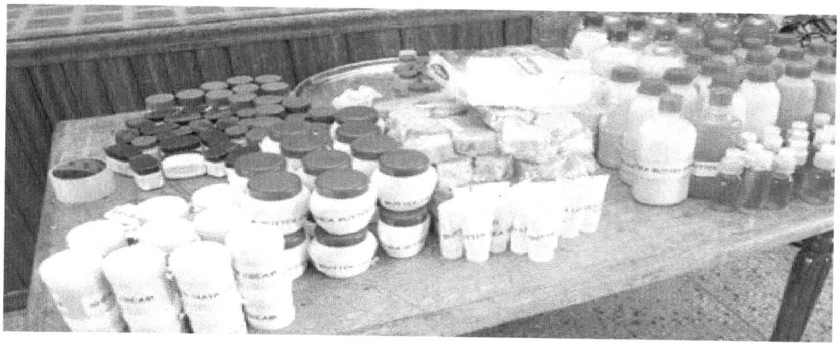

Figure 21(c): Shea-Based Products Developed at the Workshop

5. CONCLUSION

In conclusion, I wish to encourage the Participants at this forum on "Policy-Making" in Popularization of Science, Technology and Innovation (STI), to further collaborate with one another to promote Science Technology and Innovation in our various countries accordingly.

REFERENCES

[1] Nigeria Federal Ministry on Science and Technology (2017). On the updated Edition on Nigeria Science, Technology and Innovation (STI) Policy Document.

[2] Raw Materials Research and Development Council (RMRDC) Journals, Publication and Reports.

Policy-Making in the Area of Popularization of Science: Developing a Rationale Applying the Theory of Public Goods

Vahid Ahmadi

Department of Electrical Engineering, Tarbiat Modares University, Advisor to the Minister of Science, Technology and Innovation and the President for National Research Institute for Science Policy, Tehran, Iran

E-mail: v_ahmadi@modares.ac.ir

ABSTRACT: Since the 1950s, public goods theory has been providing a cornerstone for identifying and outlining the role of the public sector (government) in nation-states. Inspired by a pair of revolutionary articles by Paul Samuelson, published in 1954 and 1955, the policy scholars have accepted a rigorous definition of the term "public good", and has used the concept of public good to outline a role for government. Public goods theory shows why goods and services with some rigorously defined characteristics of publicness cannot be produced efficiently by the private sector, creating a "market failure", which implies a role for government in the production of those goods and services for which the market fails. In this paper, we apply the theory of public goods to look into the issue of "policy-making for the popularization of science." We try to develop a rationale for the government to intervene in some parts of the popularization process and not to interfere in other parts, abandoning the latter to the natural mechanisms of the private sector.

Keywords: Government, Theory of Public Policy, Public Good, Propagation of Science, Popularization of Science.

1. INTRODUCTION

When we talk about policy-making in science popularization, just like in any other field, the first question that comes up is to which extent and how the state should involve. In other words, the questions are how and in what ways the state should come to the aid of the system of science promotion in a country and what areas should be led by the natural mechanisms of science popularization and by the science promoters. These questions are fundamental in the societies whose economic and political foundations are based on a free market. We will clearly see that in the early stages of policy-making for these societies, the main concern is the role of the state, which is an essential element for those who are responsible for policy formulation. In our country, however, this question and those who are asking these questions are not taken seriously.

Usually, in our country, it is assumed that the government is responsible for all the problems as well as all the solutions. Besides, the issue of state incumbency is one of the mental presuppositions of everyone, from policymakers to ordinary people.

The reason is not so complicated. It is for decades that our political economy has been formed based on selling the raw materials, not based on taxation from a powerful private sector. That is why we have a very fragile private sector, and it is the state that is mainly an engine for our economy. As the economy is in the hands of the state, therefore all the responsibilities for the problems go to the government. Under such circumstances, when questions concerning science popularization and the role of the state and its policy-making system arise, naturally all minds tend to be inclined to plan for the role of government from top to bottom, in the process of science popularization.

This kind of approach is wrong, and the role of the government should be adjusted. In the field of science popularization for effective policy, we need a theory that tells us where the boundaries of the government are. What is the central role of the government, and how government should enter and play its role? The nature and role of government in an advanced society are defined in the public goods theory.

From the middle of the 20th century, the public goods theory provided for defining and limiting the role of the public sector (government) in the modern systems of the nation-state. Public policy theorists, inspired by the revolutionary articles of Paul Samuelson (1954) in the middle of the last century, achieved a precise definition of the term "public goods" and used this concept to define the fundamental role of government in society. Public goods theory shows why the private sector cannot produce and present some goods and services, due to their specific characteristics, optimally and efficiently. These goods and services are subject to "market failure", which means that common mechanisms in the free market fail to produce them in optimal level, and therefore, the government is responsible for producing and delivering them. In simple terms, in modern society, goods and services can be divided into two categories: ordinary goods, which are optimally produced, supplied and consumed by the free market or public goods that, due to the market failure in their production and supply, need government intervention. Thus, the role of government and sovereignty means the production and delivery of public goods and services (Morrell, 2009).

But let's be more precise and ask: what are the public goods from an economic point of view? To answer this question, we should note that when economists want to analyze a phenomenon, such as the public goods phenomenon, they consider the framework of the "market" in their mind and try to define the phenomenon in the context of this framework. At the core of the market framework, there are two main actors: Consumers and corporations (firms).

Consumers, in this view, try to maximize their utility by making reasonable decisions in their portfolio of consumer goods. Consumers sell their labor to firms, and they buy goods and services that firms offer. On the other hand, firms are trying to maximize their profits in a competitive market. The market mechanism for economists is, in most cases, the best and most effective mechanism available to achieve the optimal allocation of resources. However, the important thing is that this mechanism does not work well in some exceptional cases.

The first attempt by economists to find a position for public goods and services in the market framework has been made in the concept of market failures. The failure of the market occurs when the free market system does not work properly for any reason. One of the most striking effects of a market failure is the failure of this system to produce goods that are referred to as public goods.

Public goods such as security, national defense or public health are goods that, by definition, have two main characteristics: they are primarily non-rivalrous, which means that their use by a member of the economy does not affect the consumption of those goods by others at the same time. In other words, all people can consume these goods simultaneously and without interruption, and there is no competition for the use of these types of goods. For example, consider "security", when created and produced, security is beneficial for all people and there is no competition for it.

The second characteristic of public goods is that they are non-excludable. Non-excludable means that these goods can't easily be prevented from being consumed when they are produced. For example, "clean air" is a non-excludable good, because it's not possible to exclude some people from using it.

The result of these two features is that public goods are not naturally produced by market mechanisms at the optimal level. In other words, these essential properties make the private sector and the natural market mechanisms tend to be unlikely to produce public goods to the socially optimal level. The non-rivalrous and non-excludable nature of these goods causes that their producer not being able to take all the benefits of producing such goods (Nelson, 1959; Arrow, 1962).

The basic idea is that we use the public goods theory to reconsider the category of "policy for science popularization" to provide a rationale for government to involve into parts of the process of promoting science and avoiding to involve into other parts and leaving those parts to the natural private-sector mechanisms (i.e. universities, research institutes, and science promoters). This, of course, needs a deep understanding of the process of science popularization and separating the parts that their nature is public goods. The emphasis has to be on the creation of "infrastructures" to promote science and provide "incentives" for the actors of science popularization; actors who, in the presence of these incentives, would have the best performance in science popularization.

At the National Research Institute for Science Policy, we have had such an approach in policy-making in science popularization. The orientation of the research department of science popularization at the National Research Institute of Science Policy has been to establish a base for actors in the promotion of science, to coordinate collaboration between individuals and institutions which are promoting science. Our unit for Promotion of Science has had strong links with the Iranian Association for the Popularization of Science, and an example of joint activities with this association is the annual Science

Promotion Award. This award is held annually to appreciate the efforts of the science promotion actors and to encourage the public to work to promote the scientific insight and scientific thoughts to society and to appreciate individuals or organizations that have been the best and most effective actors in popularizing science and scientific thought in Iran.

Following up on the activities of the National Research Institute for Science Policy (NRISP) in the field of policymaking for the science popularization, we held this internationally recognized workshop, with the participation of the member countries of the Non-Aligned and other developing countries. Among the topics of the workshop, there is the title of "identifying the educational backgrounds and strategies for the promotion of science", "the developing science and technology infrastructure for the popularization of science", "the popularization of science through indigenous cultural traditions", "the role of formal and informal education in promoting science", "analyzing different approaches to education through identifying the challenges and limits of the promotion of science". The workshop will be attended by international experts presented their scholarly experiences and findings.

The workshop was held for the first time at the level of countries of the Non-Aligned Movement and the Islamic Republic of Iran. We hope that the follow up activities will bring about the networking among the experts of these countries and attract the attention of politicians, policymakers, scholars and the public to the crucial issue of science popularization and its role in All-inclusive development of developing countries.

REFERENCES

[1] Arrow, K. (1962). Economic Welfare and the Allocation of Resources for Invention. In *The Rate and Direction of Inventive Activity: Economic and Social Factors* (pp. 219–236). London: Princeton University Press.

[2] Morrell, K. (2009). Governance and the public good. *Public Administration, 87*(3), 538–56.

[3] Nelson, R.R. (1959, June). The Simple Economics of Basic Scientific Research. *Journal of Political Economy, 67*, 297–306.

[4] Samuelson, P.A. (1954). The Pure Theory of Public Expenditure. *Review of Economics and Statistics, 36*(4), 387–89.

Policy Making in Popularization of Science in Iran

Akram Ghadimi[1] **and Elaheh Hejazi**[2]
[1]Department of Political Science, National Research Institute for Science Policy (NRISP), Tehran
[2]Department of Psychology, Tehran University, Tehran
E-mail: [1]ghadimi2005@gmail.com

ABSTRACT: The popularization of science in the modern sense is a new topic in which all sciences are somehow linked. Popularization of science is a bridge between scientific literature and the discourse of the public. In simple terms, it is a tool for connecting society and the world of science. The concept of popularization of science is an attempt to reflect on scientific ideas in such a way that all individuals can understand the basic concepts of science.

In the public area, popularization of science is defined as a systematic activity with the participation of various educational, administrative (government), economic (such as the private sector and the market) institutions, and civil society that have emerged over the past decades. In fact, since the early 1980s, Western governments have comprehensively paid attention to this issue and provided a suitable environment for the popularization of science. They mobilized other actors (i.e., the scientific community, the market, and civil society) for achieving the key and important goals of this program.

The purpose of this paper is to determine the position of policymaking in the field of popularizing science and formulating national macro-plans based on the policy in this area. This research seeks to answer this question: Why is science-policymaking important in the area of popularization of science?

This research is carried out with a descriptive-analytic method. Increasing cooperation at the national and international levels, empowerment of human resources, and the provision of facilities for the development of popularization activities are among the main policies in the field of popularization of science.

Keywords: Popularization of Science, Policy Making, Macro Planning.

1. INTRODUCTION

Essential policymaking issues in each country are specific to the country. Policymaking in the popularization of science, like in any other field, requires attention to the time and place that has led to the understanding and application of science in people's everyday lives. Therefore, it is not reasonable to prescribe science popularization policies effective worldwide.

However, it is possible to find common experiences and learn from each other at times. New policy making in this area is also based on learning. The goal is not to imitate the other or to integrate the knowledge of others, but rather to reflect on the knowledge and experience of others, such as speaking the language and structure of language but speaking in ones own way. Our speech is not an imitation or synthesis of what others are saying, it is new (Peivasteh, 2019:44).

The popularization of science in the public area as a systematic activity with the participation of the various educational, executive (government, economic such as private and market) institutions and civil society is a new development that has not lasted for decades. In fact, since the early 1980s, Western governments have focused their attention on this phenomenon, providing the appropriate ground for the growth of this activity, and other key players in society (i.e., the scientific community, the market, and society). They have taken part in the movement and mobilized with them to achieve the goals of this activity (Paya, 2008).

Given that policymaking can guide the future direction of societies in different areas, science popularization policy-making can map countries' paths in the field of science and society linkage and provide the essentials for countries to achieve sustainable development. The purpose of this study is to highlight the importance of policymaking in the popularization of science. In this paper, a descriptive-analytic method is used.

Given that the study focusses on two issues of policymaking and popularization of science, the primary aim is to define and discuss these topics and then move to the process of popularization of science in Iran.

2. DEFINITION AND BACKGROUND OF SCIENTIFIC POLICYMAKING

Policymaking studies specific problems in society. The policymaking process begins with identification of the problem and deals with the various economic, political, social, cultural, and administrative aspects of the particular problem. From a practical point of view, when a policy can play its role as a guide for justifying and interpreting tasks and how they perform, it also outlines the general lines and areas of emphasis of the public sector (Compston, 2004).

Scientific policymaking is a set of principles that specify general methods of action in the scientific life of a society or a nation (Barzegar, 1992). The scientific policy comprises parts of the life of a society whose application of science and its methods make it possible for us to further our scientific principles to expand the products and services needed for our society.

Science and technology policy-making is referred to as a set of factors that determine the orientation of the science and technology system and provide the basis for achieving the goals set out in national macro documents such as the Constitution and other supranational laws.

Before the emergence of public policymaking, political studies had limited to normative and ethical areas of government and political institutions. The gap between theory and practice has led researchers to seek ways to adapt and integrate political theory and policy practice, and thorough empirical analysis of governments' current policies, to fill the gap between abstract issues and objective needs. Although the origins of policymaking science go back to the works of Lasswell[1] and Lerner[2] in the early 1950s, the institutionalization and expansion of this discipline in Western universities date back to fifteen to twenty years later, in the mid-sixties and especially in the late decade. This was due to two main reasons: first, the increasing political and social pressures for the application of the social sciences to solve everyday problems such as poverty, racial discrimination, environmental pollution, and second, the emergence of the theory regarding the role

[1]Harold Dwight Lasswell.

[2]Daniel Lerner.

of socio-economic factors (such as income, education, unemployment, and so on) as more influential factors in traditional decision-making and policymaking than traditional factors such as party tendencies and political ideologies.

Formal policymaking for science in the West (especially Germany) starts in the nineteenth century. Even if we were to look for deeper indications of the support of government for science, we could mention the formation of a royal council of experts established in 1675. The council was first based in Greenwich and now in Cambridge. The motivation for European governments' support for scientific institutions returned to economic and political interests and the expansion of their empire in the late 19th century. Governments of that time used geographical and geological research both for the benefit of newly colonized resources and, albeit with more limited but equally important incentives, for health and food production. These incentives led to the creation of powerful centers of scientific activity, health laboratories, and public agricultural research in European countries.

A broader shift in the state's support for the development of science immediately began after World War I. Scientific politics gradually became part of politics and its related issues were discussed in the dialogue of politicians. For example, Polanyi has been associated with their Marxist and liberalist views on "centralized science policy" and "decentralized politics in science". Polanyi uses the purely political concept of the "Republic of Science" to describe the scientific community and Merton's discourse of the compatibility of the democratic order and the institution of science. The Eastern Bloc establishes science academies to focus on science management (Sarukhani and Maleki, 2008).

On the other hand, the science and education institutions are inextricably intertwined and one cannot speak of promoting science without regard to policy developments in the education system. The structure of the education system of most Western countries in the 20th century was inspired by European models. This structure was traditional and did not meet the needs of the community. Since the second half of the twentieth century, we have seen dramatic changes in educational systems.

The beginning of the change in education can be traced to two major historical events: one alarmed by the global crisis in France in 1968, which arose from a deep-rooted intellectual movement that questioned the existing education system. Known as the "May Event," the event was the result of the individual, social, and intellectual work of universities, high schools, and all those who were in some way dependent on education. This was, in fact, the uprising that the young generation of France launched against the educational system.

Major discontent arose from the fact that university professors had given various proposals to the education ministers in 1959, which had not been approved. Subsequently, the students issued leaflets entitled "High School Students Have a Word to Say" and urged them to participate in school administration decision-making, planning, and activities. The French state, which had been threatened with collapse, after establishing law and order, admitted that the focus on education had been a barrier to progress, so it continued on its way regardless of existing needs and international competition. Then new laws were introduced to reform education (Raouf, 2000, p. 73). The second event was the meeting of the US National Commission for the Promotion of Education in 1981, which led to a shocking report published in 1983 entitled "A Nation at Risk" (Raouf, 2000, p. 10).

One reason for this tendency in policy-making can be attributed to the Western governments' quest for some sort of "scientific legitimacy" for their increasing interference in the management of public affairs (Ashtarian, 1997).

3. DEFINITION AND BACKGROUND OF POPULARIZATION OF SCIENCE

Some literature use generalization instead of the word popularization. There is no consensus on the use of words popularization, generalize, or generalize science, etc., but obviously, all these words have one purpose.

In the late 18th century, the concept of "science popularization" or "public understanding of science" came into being. But it was only in the 1940s that it was applied to science and technology (Ren & Zhai, 2011: 13). Popularization of science today means new topics to which all sciences are related. But, given the differences in the ways science is promoted in different areas, there is no comprehensive definition of this concept. However, the definition of the popularization of science cannot be ignored, as one of the key tools for the popularization of science is the creation of a common language or, in other words, a definition acceptable to scholars. This does not mean that there has been no definitive promotion of science, but that there is no one method that is universally accepted and credible. Therefore, the definition of science popularization for some experts and intellectuals is the transfer of knowledge and awareness of an expert to the non-specialist community.

John Goodfield believes that linking science to the public "encompasses a range of considerations, from culture, education, government to politics." Public science tends to introduce science as a matter of participation, social responsibility, and citizenship. In this regard, popularization or publicizing science and technology at the socio-economic level encourages people to choose scientific research, technology development, and intellectual work as a job. It also enables creativity and innovation, expands social opportunities, and strengthens the education system. Nowadays, science and technology have become increasingly important in all areas of social life and the popularization of science has become a strategic issue.

Ludwik Fleck emphasizes the importance of publicizing science for the advancement of specialized knowledge. To connect scientists and the public, new forms of interaction have emerged that have created great potential for knowledge production. The popularization of science is important to increase understanding of the process and progress of science, critical thinking, citizenship, and social responsibility (Schall, 2000, pp. 71–77). Bensaude-Vincent believes that in the 18th century there was no clear boundary between professional scholars and nonprofits.

Bazerman and Atkinson's studies emphasize the vital role of the nonprofit community in scientific development (Bensaude-Vincent, 2001, pp. 265–279) (Vincent, 2003: 265–279). Non-professional intellectuals and scholars were separated into distinct social groups in the late 18th and early 19th centuries. This separation, on the one hand, created a boundary between the self and the non-self, on the other hand, triggered a wave of "make science accessible to all" by the general public. These two trends gradually created public science and academic science in two parallel networks, creating the need for popularization of science as a bridge between scholars and society. However, in the 19th century, cooperation between these two parts was not easy. In the twentieth century, standardization of science was achieved through rigorous and professional training, regular practice, and regular scientific communities. At this time, the ability of nonprofits to participate in scientific processes was further restricted by specialized and scientific language, expensive equipment, and complex and subjective theories such as modern physics.

After this time, the scientific texts available to the public were not of sufficient quality, and the gap between scholars and the general public increased. As Bensaude-Vincent writes: "From then on, any non-professional activity in the field of science was regarded as quasiscience." To understand more about the popularization of science, we first need to examine the concept of science more closely (Ghadimi, 2017: 18). In the book *Science and Human Values*, Bronowski defines science as follows: "the organization of our cognition and knowledge in a way that most of it are capable of deciding" Bronowski, 1956, p. 7). Karl Popper believes that "science is a biological phenomenon ... that results from conscious cognition" (Popper, 1971:5–6). Bensaude-Vincent defines science as: "Science is rooted in common sense and it is nothing but a complex and expansive form of public and popular reasoning" (Vincent, 2001: 108).

Drake Gilbert Ross believes "Scientists are practitioners of science who contribute to the production and creation of scientific knowledge through scientific methods and experimental research. They present scientific knowledge through accepted components and align it with accepted language patterns and guidelines provided by the scientific community" (Ross, 2004). This definition represents the "popularization of science" and the presentation of scientific knowledge in a form and style that is understandable to the public. The "generalization" part of science encompasses this aspect. Historical evidence suggests that the nineteenth and twentieth centuries have been the most popular popularization of science. "Scientists produce science while ... for a variety of reasons, they also produce, read, and participate in the generalization of science," says Dant Paul. Their participation, whether as a contributor, producer, or reader of science in the generalization of science, will have consequences for science.

In the journal *Current Science*, popularization of science has been defined, in fact, as popular science. This definition expresses the true purpose of the popularization of science (*Current Science*, 1993).The definition of the popularization of science from the perspective of Paul and others includes the following:

1. The popularization of science is a continuous affair;
2. Writing journal articles in newspapers is official popularization;
3. Textbooks are a form of popularization of science;
4. Children's books, like journal articles, should have a high level of scientific understanding.

Dr. Abdul Kalam believes "In a world surrounded by science and technology, the popularization of science is of particular importance because it makes most people aware of the consequences of science and technology."

According to the above definitions, there are two broad meanings of the popularization of science:

1. The process of disseminating specific sciences to a wide range of people;
2. Changing the public view and changing the culture of the people towards the use of science and its benefits.

We explain some other definitions of the concept of popularization of science:

1. Popularization of science is a bridge between scientific literature and the discourse of ordinary people;
2. In simple terms, popularization of science is a tool for linking and communicating between the society and the world of science;

The concept of popularization of science is an attempt to reflect scientific ideas in a way that all people can understand the basic concepts of a science, generally called the public understanding of science, popularization science (Communicating and popularizing science, 1999).

According to the Iranian Association for the Popularization of Science, "Popularization of science is any activity that is undertaken to generalize science, to spread scientific thought and thinking, or to interest people or groups of them in science and scientific activities." In other words, it is adapting and modifying sophisticated scientific ideas and findings so that non-specialist audiences can access it; romoting activities that popularize science to educate ordinary people, foster public confidence in science, and promote science. Popularization of science, in general, is an attempt to visualize scientific thought so that anyone can understand the underlying science concepts and understand the essence of it. Cloitegr and Shinn believe that in the popularization of science, language does not describe reasoning and imagery but rather tends to create a lack of conceptual understanding. Thus, "... the popularization of science is not an effective tool for transmitting cognition better than the physical world, but rather the ability of popularization of science in forming relationships that forge a scientific subject and social environment ..." (Ghadimi A., 2009a, pp. 18–20).

According to the definitions, perhaps the most important goals of popularization of science can be divided into several general categories.

1. Using science and scientific method to solve community problems;
2. Belief in the ability of science to solve societal problems;
3. Teaching people to make good use of science and technology;
4. Fighting superstition and pseudoscience.

In the developed world, with all its background of science popularization, promotional activities have been in the form of dispersed efforts by individuals and populations. However, since the 1980s, the government started to tackle this issue, and so popularization of science "became an outwardly highly influential social reality" (Ghadimi & Nazifkar, 2010: 16).

One of the new features of science popularization is the evolution of its tasks and missions. Over time, with the advancement of science and knowledge, and especially in the areas of advanced technologies, the task of popularization of science has strongly increased. In today's world, the number of people in a society who have heard at least some of the news about sophisticated technologies is by no means comparable to a few years ago. Improving the teaching of science in primary and secondary schools is another historical development in the field of science popularization (LI, 2005).

Given the above, it can be said that popularization of science in the modern sense began in the 1980s in some developed countries. The popular acceptance of science, traditionally referred to as the popularization of science, has its roots in the 17th century and is an early form of the popularization of science. Scientists and scientific societies were at the forefront of this form, and the popularization of science was a one-way process, in which information flowed from the scientifically conscious to the illiterate.

In the 19th century, the collaboration between the two departments of public science and academic science was not easy to achieve. In the twentieth century, standardization of science was achieved through rigorous and professional training, regular practice, and regular scientific communities. In the twentieth century, and especially in the 1980s, the concept became applicable today. In 1993, the United Kingdom first spoke of a strategy to increase public understanding of science and technology. Before that, in 1972, the United States had cited research on popular science literature as one of its goals in the field of engineering. Popularization of science has gradually expanded its presence from developed to less developed or underdeveloped countries. France is another country that is known as the origin of science and the popularization of science. Unlike the United Kingdom and the United States, where the first productive institutions of scientific thought were the product of a group of intellectuals or capitalists, the formation of science-producing institutions took place directly in France with the support and encouragement of the state (Paya, 2006, p. 119). The popularization of science, which was once only the publication of the latest scientific achievements in the eighteenth and nineteenth centuries, has now taken on new forms. Forms such as science shops, restaurants, and cafés, science parks and museums, and public scientific conferences all indicate the emergence of a new phase of this historical evolution (Vesali, 2007).

4. THE PROCESS OF POPULARIZATION OF SCIENCE IN IRAN

The process of popularization of science in Iran is a new and nascent process that can be traced back to the beginning of the last half-century, i.e., the start of radio in Iran. These processes in Iran, like all countries, have had many ups and downs. It began with the voluntary and dispersed activities of individuals, while its various stages, including conceptualization, institutionalization, and structuring, are hard to distinguish. For the past few years, the popularization of science has been limited to individuals and groups. But in recent years, the concept of popularization of science and the generalization of science have taken on a more realistic meaning (Ghadimi & Nazifkar, 2010).

The first sparks of the formation of serious thought about the popularization of science in Iran may be traced back to 1999, in conjunction with "A Study of the Strategies for the Development of Science in Iran", in which the idea of establishing the Iranian Science Popularization Association was put forward. It is the first official activity in Iran to publicize science or promote science. Following the idea of establishing the Science Popularization Association, further complementary views were developed on the creation of science clubs, museums of science and technology, plant responses, experimental workshops, laboratories, special seminars, and scientific workshops. It should be noted, of course, that the establishment of the Science Popularization Association is symbolically the first official step in the popularization of science in Iran, but prior to that the publication of scientific journals and journals such as *Scientist Magazine*, broadcasting scientific programs on radio, television, and the media had been there. Scientific awards, sometimes held by some ministries or universities, form volunteer groups to familiarize people with specific topics, especially health and agriculture, and most importantly the literacy movement in its own right have comprised popularization of science (Ghadimi A., 2009b).

Radio and television in Iran have been an important and effective institution in the popularization and publicizing of science since their inception. It should not be forgotten that radio and television are technological phenomena and have had an impact on Iranian society. The history of science popularization programs on television goes back to the '40s. Influenced by the space-age revolution (Sputnik Launch, the first human astronaut Gagarin and Apollo launch) and the use of radio and television media to reflect these developments in Iran, the movement toward science programs began. The news shock of the evolution of the space age led to some limited, but decentralized, unplanned efforts to introduce science to the public (Vesali, 2007).

Scientific public journals such as *Danesmand, Dansgar, Kavosh, Etelate Elmi* were important scientific-public journals in Iran. *Salamat, Machine Magazine*, and some information technology journals are typical examples of public science and technology journals, of course. Newspapers and magazines with a wide audience such as *Hamshari, Jame Jam,* and the like have special science and technology services and pages that are relevant to the public (Vesali, 2007).

5. POLICY MAKING IN POPULARIZATION OF SCIENCE IN IRAN

Public policymaking knowledge is a new subject, but governments have used it as a practical measure to solve public issues in all societies. According to the major documents of the science and technology system, one of the fundamental issues of converting science into the main discourse of society is to create a favorable environment for the production of science and technology. The most obvious cost for dragging attention to science in society is its social cost, in the sense that drawing attention to the issue of science-popularization requires the formulation of a policy in which the pillars of science popularization system are put on the agenda.

A forward-looking approach to the popularization of science requires reducing the gap between dreaming and achieving goals under national documents in a fully scientific framework. We need policymaking in this area to implement popularization of science among different sections of society. Actions and activities in the field of science popularization in the country can be successful and effective when they are politically formulated in such a way that they can be implemented and evaluated.

Popularization of science will enhance public oversight policies. Encouraging people to be sensitive to scientific and technological policy-making will strengthen public oversight, and in practice, will cause many of the wrong decisions of scientific policymakers to be tested by public opinion before being implemented.

In Iran since the Islamic Revolution, although many organizations and centers have been active in policy-making and decision-making in the field of science, there has been no unified organization at national level and there has been no harmonization between the involved institutes. However, as there is insufficient coordination between the executive mechanism of the map and the current mechanisms in the country's executive agencies, the comprehensive scientific map has failed to coordinate with the country's executive agencies. Therefore, in the first step, to reform the institutions and the structures that govern them, it is necessary to analyze and identify the obstacles to their integration. Policy formulation and policymaking are important measures of social systematization. Policymakers are defining the way forward by formulating the future horizon and outlook.

Transformation and change in any structure is one of the factors affecting its dynamics and growth. In the process of popularization of science or general scientific policies, the responsibility of the agents is one of the key issues. In other words, institutions and organizations, on the one hand, and individuals, on the other hand, as agents for the popularization of science, are capable of evolving in the promotion of science. Sometimes it can be the government that acts as the primary and most prominent agent among different social groups and classes. Despite the differences between the activities of different agents, due to the similarity of their features, they will have an effective role in the popularization of science. However, theoretical dilemmas about the status and role of agents underlie some theoretical underpinnings in the field.

What matters is the continuity and connection of the agents with each other. The theoretical reflection on the promotion of science inevitably leads us to the view that there is no escape to the pathology of the process of science popularization except the precise identification of the agents and mediators of the popularization of science and the tools used. However, the multiplicity of brokers in the field of science popularization, conflict and dichotomy are expelled in this group, and to access the normative status of brokers to stabilize the process of popularization science in the country, brokers' performance should be regarded as the highest basic concept in promoting science.

Problems, such as the lack of awareness of the need to promote science among policymakers or in other words, the lack of a proper legal and regulatory framework for science popularization activities, lack of clarity on the role of various actors and the like, indicate the lack of special policymaking in science popularization in the country. As a result, science popularization policymaking is one of the essentials for development that must be put at the top.

6. CONCLUSION

The advent of the Third Millennium and the beginning of the Information Age have created new relationships in the field of social life, with the accompanying education policy and training of individuals capable of understanding new frameworks and adapting them. Therefore, in today's world, the policy of acquiring global technology and localizing it and targeted policymaking in public education and acculturalization are two essential pillars of sustainable development (NiyazAzari, 2002).

The activities of popularization and publicizing science in Iran are very wide and varied. With a glance at promotional activities in Iran, one can see their dispersion and decentralization, and in many cases their spontaneity. At the mid-levels, there is a multiplicity and variety of promotional activities. This multiplicity, variety, and wide range of activities demonstrate the enormous potential for the development of science-promotion activities. Policy-making in popularization of science is one of the essentials of sustainable development.

The need for policy-making for popularization of science along with attracting public support also facilitates the process of budget allocation for scientific activities which will ultimately lead to further science growth. Ultimately, social development is about making citizens aware and critical, and the popularization of the scientific spirit is one of the best tools to achieve such a level of awareness.

In general, science popularization policy can lead to the provision of various essentials such as support policies for all science popularization activities, increased cooperation at national and international levels, human resources empowerment, and the provision of necessary support facilities. Scientific and research activities in the field of science popularization are needed to achieve comprehensive development. We need to converge activities and create a national reference in the field of science popularization. On this basis, it can be said that countries need to popularize science to achieve sustainable development.

REFERENCES

[1] Ashtarian, K. (1997). Policy Sciences and Achievements for Iran, *Journal of Faculty of Law and Political Science*, Tehran University, Vol. 37 [in Persian].

[2] Barzegar, A. (1992). A Look at the Concept of Scientific Policy, The Center for Scientific and Research Policy. *Rahyat, 3 [in Persian].*

[3] Bensaude-Vincent, B. (2001). A genealogy of the increasing gap between science and the public. *Public Understanding of Science, 10*(1), 99–113.

[4] Bensaude-Vincent, B. (2003). A Geneology of the Increasing Gap between Science and the Public, *Public Understanding of Science: Discourse Studies*, 5(2), Sage Publications, Pages: 265–279.

[5] Bronowski, J. (1956). *Science and Human Values.* New York: Julian Messner: Inc. Retrieved.

[6] Compston, H. (2004). *Handbook of Public Policy in Europe.* Palgrave Macmillan.

[7] Cornelis, G.C. (1998). Is Popularization of Science Possible? *The Paideia Archive: Twentieth World Congress of Philosophy, 37*, 10–15.

[8] Ghadimi, A. (2009a). *Auditing Presidential Popularization of Science.* Tehran: Iranian Association for Popularization of Science [in Persian].

[9] Ghadimi, A. (2009b). *Drafting of Science Popularization Indicators on Achieving Proper Modeling.* Tehran: National Research Institute for Science Policy [in Persian].

[10] Ghadimi, A. and Nazifkar, G. (2010). Studying the Status of Science Popularization in the Islamic Republic of Iran: Must to the Real. *Rahyaft, 47*, 13–22 [in Persian]. *Current Science* (1993). 65(6), 25 September 1993.

[11] LI, J. (2005). Study of Science Popularization in Modern Cities. *Shanghai Institute of Science.*

[12] NiyazAzari, K. (2002). Human Behavior and Relationships in Educational Organizations of the Third Millennium, Tehran: Metacognitive Thought Press [in Persian].

[13] Paya, A. (2006). *University Scientific Thinking, Innovation and Public Domain.* Tehran: Institute for Cultural and Social Studies [in Persian].

[14] Paya, A. (2008). Popularization of Science in Society, A Philosophical Assessment. *Journal of Science and Technology Policy* (1) [in Persian].

[15] Peivasteh, S. (2019). STI Policy Making: Social Aspects and Consequences. *Journal of Science & Technology Policy, 11*(2), 43–57 [in Persian].

[16] Popper, K. (1971). *The open life and its Enemies* (Vol. 2). New Jersey: Princeton.

[17] Raouf, A. (2000). Global Movement for Teacher Education Improvement. *Ministry of Education, Faculty of Education* [in Persian].

[18] Ren, F. and Zhai, J. (2014). *Communication and Popularization of Science and Technology in China.* China Science and Technology Press and Springer-Verlag, Berlin Heidelberg.

[19] Ross, D.G. (2004). *The Social Role of Popularized Science.* MA thesis, July: Virginia Polytechnic Institute and State University.

[20] Sarokhani, B. and Malek, H. (2008). A Sociological Analysis of the Role of Scientific Associations in Educational and Research Policy Making Process, Humanities Congress [in Persian].

[21] Schall, V. (2000). Science Education and Popularization of Science in the Biomedical Area: Its Role for future of Science and of Society. *Mem Inst Oswaldo Cruz, 95*, 71–77.

[22] UNESCO (1999). Communicating and Popularizing Science. *World Conference on Science.* Budapest: UNESCO.

[23] Vesali, M. (2007). *Observation and Study of National Science Popularization Policy in G8, D8, India and China.* Tehran: National Research Institute for Science Policy [in Persian].

The Role of Future-Oriented Researches in Policy Cycle of Science Popularization

Akram Ghadimi* and Sahar Kousari
Science Policy Department, National Research Institute for Science Policy (NRISP), Tehran, Iran
E-mail: *ghadimi2005@gmail.com

ABSTRACT: This chapter aims to depict future-oriented researches as extended service encounters between foresight practitioners and policy makers in the science popularization realm. The implications of this perspective for evaluating the outcomes of such programs are to be explored.

The range of activities comprising foresight process (scoping, recruiting, scanning, appraising, advising, renewing) and policy cycle (agenda setting, policy formulation, policy adoption, implementation, evaluation) is reviewed, along with the various objectives that may underpin in science popularization. The more substantial foresight programs and policy cycle are seen in terms of a series of steps, in each of which various partners can be involved in generating service outcomes and later steps of the process. The arguments are illustrated with insights drawn from various cases.

Often, of course, future-oriented researches are deliberately designed to feed in to the first three steps of the policy cycle, being triggered by the agenda setting stage of the latter, but also quite possibly helping to further define the agenda. The various steps in foresight should be contributing to further understanding of the problem, formulation and assessment of policy alternatives, and selection of preferred policies. The foresight activity could thus be seen as simply providing knowledge for the early stages of the policy process of science popularization. This can be important in that policy implementation may be dependent on, or much affected by, the engagement of stakeholders of science popularization in foresight. The "Advising" and "Renewing" stages of the Foresight cycle are liable to be contributing to policy implementation in other ways, too.

Foresight programs are frequently commissioned, and often have significant influence on policy cycle of science popularization realm. Attempts to systematically evaluate these efforts have begun, and this paper stresses the need to be aware of the complex interactive nature of foresight, highlighted by viewing it in policy cycle of science popularization.

Keywords: Foresight Programs, Policy Cycle, Science Popularization, Stakeholders.

1. INTRODUCTION

Popularization of science is a bridge between the scientific literature and the discourse of ordinary people. It is a tool for connecting and communicating between the scientific community and the world of science; the concept of popularization of science is to strive to reflect scientific thought in a way that all people can understand the basic concepts of a science (Fujun & Jiequan, 2014).

According to the Iranian Association for Popularization of Science, "Popularization of science is any activity that aims to generalize science and expand scientific thinking." According to the definitions, the most important goals of popularization of science can be divided into several general categories:

- Using science and scientific method to solve community problems;
- Belief in the ability of science to solve societal problems;
- Teaching people to make good use of science and technology;
- Fighting superstition and pseudoscience.

But achieving these goals requires future-oriented policy-making to make the right decisions in the field of science popularization by predicting the future effects of the policies adopted. In the following, the process of science promotion policy making, the implications of a forward-looking approach and the implications of future studies in the field of science promotion policy making are studied.

1.1 The Policy Making Process of Science Popularization

The policy-making process in the field of science popularization consists of five stages. Each of these steps and their examples in the field of science popularization are described below.

First Stage: Agenda setting of science popularization.

A policy is set up to solve a problem or eliminate a need. Therefore, it is necessary to draw attention to it and to understand the problematic nature of the current situation. Prioritizing one issue over other important issues depends on getting more attention among other issues. In this regard, focal events can present current and potential future problems to the public and different stakeholders and prioritize policymaking through several indicators (Yazdani, 2018).

When a problem is identified, for example in the field of science popularization, issues such as rule of law, lack of public awareness of the individual, group and national players, public demand for the subject of science popularization and decision makers are on the agenda; the next steps are policy design and decision making through the policy process.

Second Stage: Designing and formulating policies in the field of science popularization.

Policy design involves actions that policy makers provide with appropriate analysis of the options available and the assessment of social space around the issue on the other. At this stage, basic criteria such as content, economic, political and executive feasibility determine the basic criteria. Examples of alternative policy priorities in the field of science popularization are the establishment of a national institute, the establishment of science popularization offices in universities, the establishment of a ministerial office in the Ministry of Science, Research and Technology, etc.

Third Stage: Decision making in science popularization.

This stage of the policymaking process has a number of complexities that policymakers must consider when making the final policy option. Different goals have conflicts, ambiguities exist and policies are often designed without causal theory, so it is difficult for policymakers to understand which particular type of policy will be most effective.

The decision-maker attempts to come up with an option that will bring the most gains to the targeted policy community and the decision-maker, with little or no scrutiny of the realization of public policy and the consequences that will occur after implementation of this policy. For example, the most efficient decision to develop science popularization ecosystem is the creation of an independent institution that can create value and contribute to the generalization of science.

Fourth Stage: Implementing science popularization policies.

After decision-makers have identified the preferred option among the available policy options, different actors in the policy process seek to play their role in the public policy implementation. Accordingly, policy implementation is one of the most difficult issues for policymakers and policy failure is one of the most frustrating parts (Fujun & Jiequan, 2014). If science policies in popularization are implemented well, one should hope for the following results:

- Production, creativity and acceptance of diversity
- Accept competition rather than monopoly
- Growth with distribution
- Promoting community health and welfare
- Scientific choices rather than traditional ones
- New perspectives for the economic arena
- Opening new cognitive channels
- Improving the quality of life.

Fifth Stage: Evaluating science popularization policies.

Evaluation shows the actual performance or results compared to the predicted goals and expected results. Policy evaluation can be categorized into three categories: administrative evaluation, judicial evaluation, and policy evaluation.

1.2 Functions and Impacts of Future Studies

The functions of futures studies are discussed in various ways (Fujun & Jiequan, 2014; Yazdani, 2018; van Dorsser, Walker, Taneja & Marchau, 2018). All of these functions can be summarized in the three main functions of policy informing, policy consulting, and policy implementation facilitation (van Dorsser, Walker, Taneja & Marchau, 2018; Walker, Lempert & Kwakkel, 2016), and the following unwanted or requested effects are listed in Table 1.

Table 1: Functions of Future Studies in Science Promotion Policy Making[4]

Targeted and/or Unwanted Impacts	*Function*
• Increased understanding of the scope of science popularization • Personal learning: Awareness of options for promoting science among actors, generating debate, clearer context and other stakeholder views • Expanding futures research skills in broader circles of science popularization • New combination of experts and stakeholders, common understanding (knowledge network) among actors in science popularization	Informing
• Understanding and continuing the common understanding among actors promoting science	
• Integrating new empowered actors and their perspectives and incorporating these perspectives into a society in which a new problem is formed	
• Reveal hidden programs and goals of promoting science • Effective actions in the field of science popularization policies	Policy consulting
• Provide policy recommendations and alternatives for implementation • Enable and support learning processes in the field of science popularization • Identify hidden barriers to introducing transparent and open collaborative processes	
• Impact on actors' research/policy programs, both in the private and public sectors • Formulating and implementing new policies in the field of science popularization • Using Foresight Elements in science popularization and Planning Procedures	
• Creating collective learning processes in the field of science promotion • Interpreting future shared perspectives, creating long-term perspectives for science popularization • Awareness of system features	Facilitate implementation
• Formation of implementation networks • Create follow up activities	
• Improving the coherence of science popularization policies • Cultural changes of long-term intuitive systematic thinking in the field of science popularization	

2. CONCLUSION

Integrating and adapting future research studies and policy making in the field of science popularization can increase the quality of forecasts in the long term and lead to improved policy making. Therefore, considering the functions of future studies in the policy making process at any stage can help policymakers. Table 2 presents future research functions that help policymakers at different stages of the policy cycle in the science popularization realm.

Table 2: Future Research Implications in Different Stages of the Policy Cycle of Science Popularization (self-compilation)

Future Studies Functions at this Stage of the Policy Cycle	
• Discover issues early • Focus on long-term issue • Emerging topics • Weak signal • Wild cards	Defining the problem in the field of science popularization
• Alternative thinking • Revealing hidden plans and goals • Providing alternative policy recommendations • Formulate new policies	Alternative solutions
• Consensus • Creating common awareness and understanding • Integrating new empowered actors and their perspectives	Select preferred alternative
• Stakeholder engagement to create commitment to implementation • Communication with the media • Synchronizing and engaging with the policy cycle	Implementation
• Internal observation capability • Better policymaking • Increased transparency • Increase the level of legality	Evaluation

Finally, future oriented research can lead to better policy making in the field of science popularization, but linking futures studies to policy making is one of the biggest challenges facing the futurists. This chapter attempts to explain the impact of future studies on the policy-making cycle of science popularization.

REFERENCES

[1] Fujun, R. and Jiequan, Z. (2014). Communication and Popularization of Science and Technology in China. Springer.

[2] van Dorsser, C., Walker, W.E., Taneja, P. and Marchau, V.A. (2018). Improving the link between the futures field and policymaking. Futures, 104, 75–84.

[3] Walker, W.E., Lempert, R.J. and Kwakkel, J.H. (2016). Deep Uncertainty. Boston: MA: Springer.

[4] Yazdani, M. (2018). Understanding the Public Policy Process; An Overview of Concepts, Theories and Models. Public Policy Research Quarterly, 189–201.

Section 2

STI Popularization Reports

Country's Status Report on Policy Making in Popularization of Science, Technology and Innovation (STI), Bhutan

B. Dawa Tshering
Higher Education Planning Division, Ministry of Education,
Royal Government of Bhutan, Thimphu
E-mail: dawatshogyal04@gmail.com; dawtshering@moe.bov.bt

ABSTRACT: A country that does not give importance to advancement in science and Technology has a very dim opportunity to grow and develop. Thus, in this 21st century, every nation strives to promote science, technology, and innovation. Bhutan having realized its immense importance, her education system gives top priority in the popularization of Science, Technology, and Innovation (STI) at all levels. This paper provides a brief history of the Bhutanese education system and its paradigm shift on the popularization of Science, Technology, and Innovation (STI).

Keywords: STEM, Popularization, Implementation.

1. INTRODUCTION

Bhutan is a small landlocked country nestled in the Eastern Himalayas. It is sandwiched between the two giants of the world, China in the north and India in the south. Bhutan is the second least populous country in Southeast Asia after the Maldives with a population of about 0.75 million people Thimpuis its capital city.

Gross National Happiness (GNH) is a philosophy that guides the government of Bhutan. His Majesty the Fourth King of Bhutan developed the GNH. It's a holistic and sustainable approach to development, which balances material and non-material values with the conviction that human wants to search for happiness. The GNH is translated to life in Bhutan through four pillars: Conservation of the Environment, Equitable and Sustainable Development, Good Governance, and Preservation of Culture (Ghalley, 2019).

The Sustainable and Equitable Socio-Economic Development is an important element to the happiness and wellbeing of the people. The education in Bhutan plays an important role in promoting this pillar so education is placed as a priority list in every development plan in Bhutan. This paper is a desk review of how education is been prioritized with the changing needs of the world.

2. EDUCATION IN BHUTAN

Although the first modern school in Bhutan was opened in 1913, access to modern education remained limited to only a handful of people. The number of schools increased in the 1950s, but it was still common for the Bhutanese people to opt for monastic education since educating the children in the modern lines was new and not extremely popular.

When the first Five Year Plan was launched in 1961, there was a massive paradigm shift. The number of schools was expanded to address the basic educational needs and develop human resources required for the socio-economic development of the country (Royal Government of Bhutan, The Royal University of Bhutan Strategic Plan 2013–2020, 2015). Today, there are 512 schools and 79 extended classrooms across the country with 167,108 students and 9574 teachers as of 2018. The adjusted net enrolment was 96.8% with the teacher-student ratio of 1:17.45 (Royal Government of Bhutan, Annual Education Statistics, 2018).

The Government provides free and compulsory education to all children of school-going age up to 10th standard and ensures that technical and professional services are made generally available, and higher education is equally accessible to all based on merit (Namgay, 2017).

The first college was established in the year 1983 followed by various colleges spread across the country. This eventually led to the establishment of the Royal University of Bhutan (RUB) in 2003 and Khesar Gyalpo University of Medical Sciences of Bhutan (KGUMSB) in 2015 (Ura & Zangmo, 2012). The establishment of the two Universities brought all tertiary education institutions under them, which were previously governed by the relevant ministries and agencies. Presently 19 tertiary institutions are providing various programs. The total number of students studying in various TEIs within the country for the academic year 2018 was 11,980 (6337 male and 5643 female) and the total number of students within and outside Bhutan was 17818 which includes scholarship and privately funded students (9531 male and 8287 female). The Gross Enrolment Ratio (GER) at the tertiary level within & outside Bhutan is 24.54% (Ura & Zangmo, 2012).

3. THE POPULARIZATION OF SCIENCE, TECHNOLOGY, AND INNOVATION IN BHUTANESE EDUCATION

With access to education firmly in place, attention is now turned towards improving the quality of education. The Ministry of Education plans to bring reforms to develop a differentiated curriculum, expand and diversify the Technical and Vocational Education Training (TVET) programs and provide a greater focus on Science, Technology, Engineering and Mathematics (STEM), which is expected to create a greater range of education and training pathways in schools. The Figure 1 represents the differential pathways in schools.

To meet the demand of the growing economy in scientific and technological innovation, and to provide a focused environment for students with an aptitude and competencies in science and technologies, the 19th National Education Conference held in December 2018 endorsed the establishment of one Premier School for STEM Education. An in-depth study to develop an action plan for the establishment of a Premier School for STEM Education within 2019 (State of Tertiary Education, 2018).

The long term aspiration of the Ministry is to establish three premier schools in STEM spread across three regions in the country, west, central and east. However, within the 12th FYP period, the Ministry proposes to establish one premier school in STEM in one of the three regions as it will entail infrastructure, resources and time.

To cater to the needs of different abilities and interests, and provide greater flexibility and alternative pathways for students, the 19th National Education Conference endorsed that a Differentiated Curriculum in science and mathematics for class IX be implemented

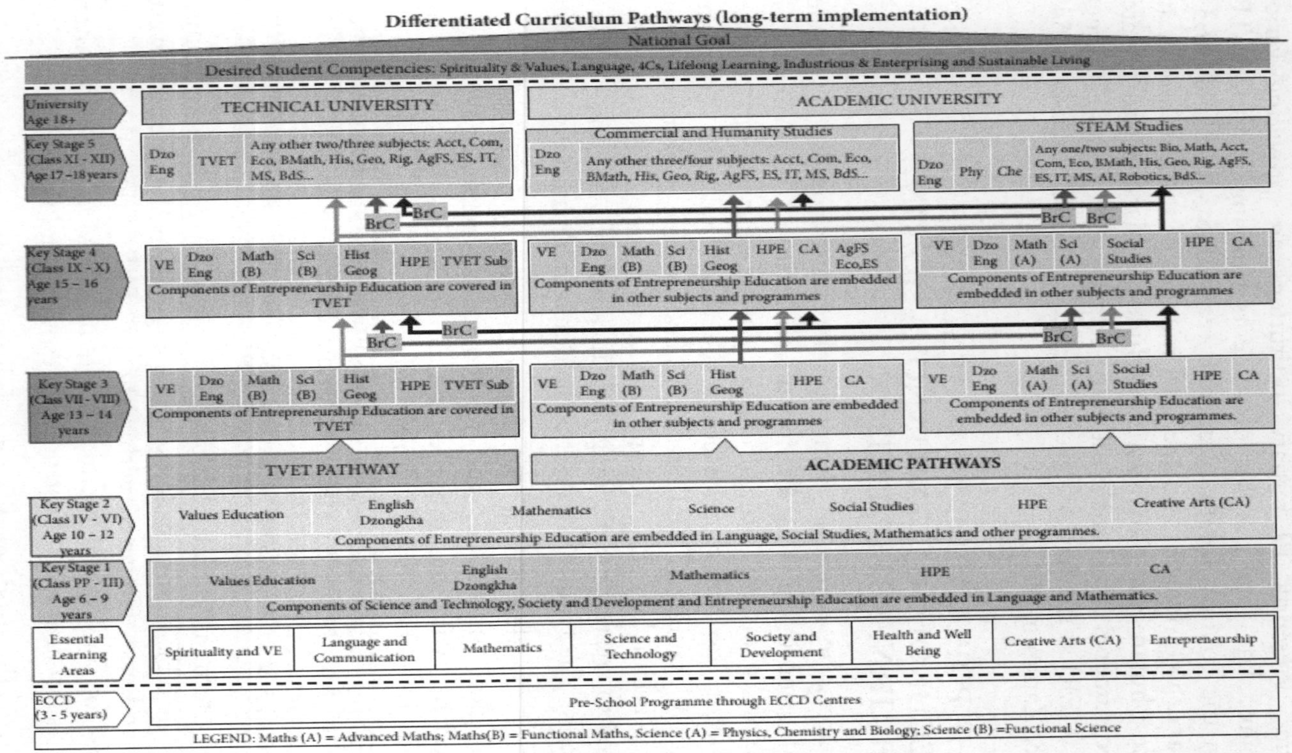

Figure 1: Differentiated Curriculum Pathways in Schools

(*Source:* School Planning and Construction Division)

from 2020. The Conference also endorsed that the Royal Education Council should prepare a plan for the piloting of differentiated curriculum in science and mathematics, in collaboration with other relevant agencies and present to the Curriculum Technical Advisory Board and the Council for endorsement (Wangmo, 2019).

Further, to encourage scientific innovation and creativity among school children, the STEM Olympiad is held every year at the national-level since 2017. It is organized by the Ministry of Education (MoE) in collaboration with the Royal Education Council (REC). The STEM Olympiad provides a platform to showcase the talent, innovation, and creativity among students (Dorji, 2014).

The government has, in recent years, invested heavily in providing ICT infrastructure to schools and communities across the country (Bhutan. Ministry of Education, 2017). The ICT classes have been introduced from class IV to XII on a phase-wise basis since 2017. Further, to harness the potential and benefit of Information and Communication Technology in the teaching and learning process, the Education Ministry launched its second ICT Master Plan called iSherig-2 on 23 May 2019.

iSherig-2 is one of the strategies in the 12th Five Year Plan to take ICT in education further as a means to improve teaching and learning. The total budget estimated for the implementation of the master plan is 4.4 million USD. The ministry through this master plan aims to motivate lifelong learning and make teaching and learning effectively. It will also enhance the use of ICT-integrated teaching and learning resources, build the competency of educators, learners and supporting staff and strengthen ICT infrastructure and connectivity for better learning and educational services. And the ministry intends to achieve all these through six programs and 21 projects that will be implemented by all relevant agencies. The master plan is expected to guide the government's ongoing investment in the education sector over the next five years. It is also expected to make learners globally competent through the use of emerging and relevant technology. iSherig-2 aims to achieve the above-desired outcomes through the programs and projects under three key areas of iAble, iBuild and iConnect (Royal Governance of Bhutan, 2010). The overview of iSherig-2 is summarized in the Figure 2.

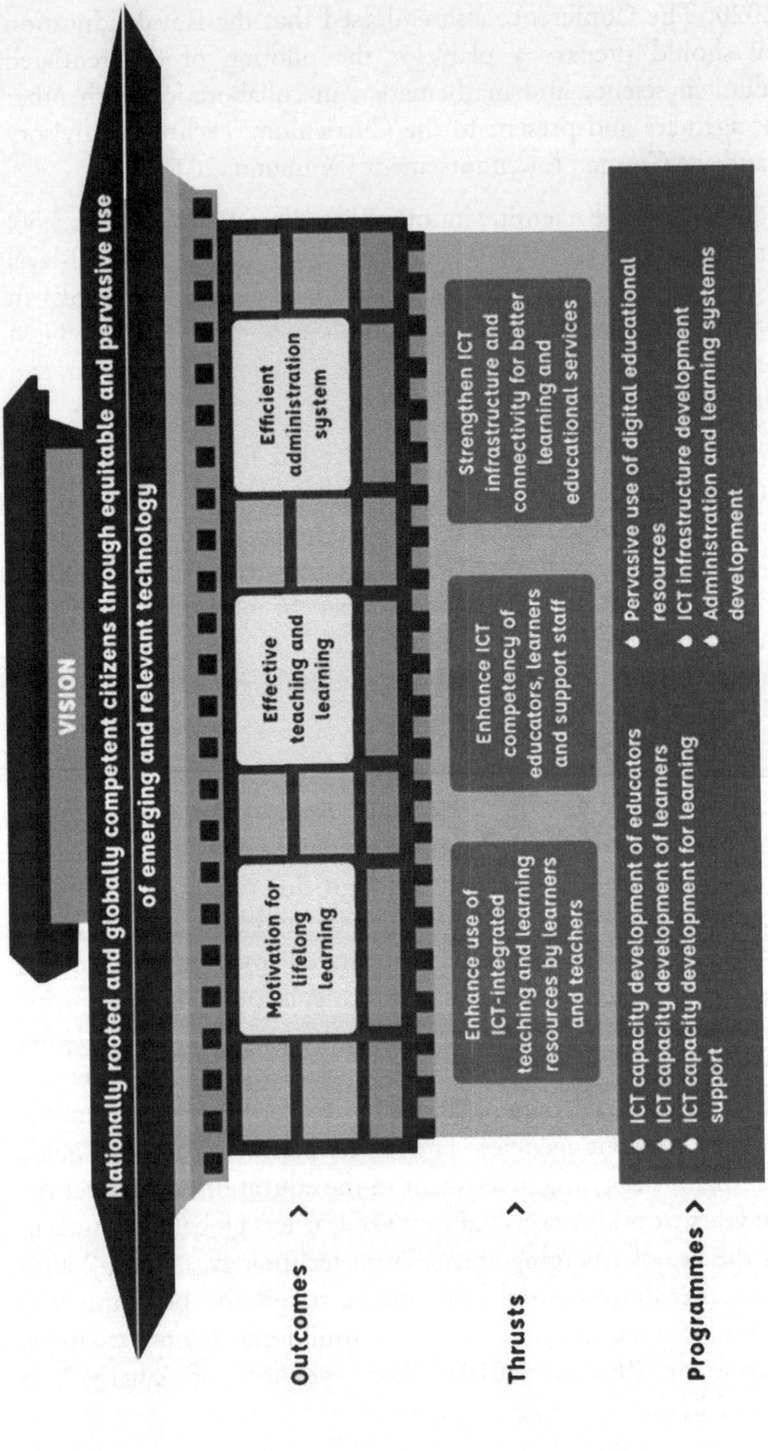

Figure 2: Overview of iSherig-2

(*Source:* iShering-2, Education ICT Master Plan 2019–2023)

At the tertiary level, the government aspires to increase the STEM programs from 35%–50% within the period of 10 years (from 2017–2027) (Palden, 2018). Several measures have been planned and introduced to achieve the target.

To develop human resources in ICT, the government established a new College, Gyelpozhing College of Information Technology (GCIT) in 2017, with the vision to be the 'Center of Excellence in Information Technology steeped in Gross National Happiness values.' (Bhutan. Ministry of Education, 2017).

The new college, GCIT which is presently offering only Bachelor of Computer Application will be soon introducing Bachelor of Science in Information Technology. The College also plans to diversify the programs by offering courses in animation, multimedia, and information security within a few years (Dema, 2018).

The other colleges that are dedicated to offering STEM programs such as the College of Science and Technology (CST), Jigme Namgyal Engineering College (JNEC), College of Natural Resources (CNR) are also working towards increasing the number of science related programs. For instance, this year CST introduced BE in Instrumentation and Control Engineering, and BE in Engineering Geology. Likewise, JNEC has introduced two Bachelors's degree programs in BE in Mechanical Engineering and BE in Survey and Geo-Informatics. College of CNR has introduced B.Sc. in Organic Agriculture and B.Sc. in Food Science and Technology. These Colleges have plans to increase the number of STEM programs so that there will be an increased opportunity for the Bhutanese students to pursue tertiary education in STEM areas.

Since Bhutan is a developing country, Science and Technology are seen as a crucial tool for further growth and development. Failure to invest and popularize science and technology will thwart its progress. Thus, realizing the importance of Science and Technology in the development of the nation, Bhutan is committed to giving utmost importance to the popularization of Science, Technology, and Innovation with the proper policy in place.

REFERENCES

[1] Bhutan. Ministry of Education (2017). *Tertiary Education Roadmap for Bhutan:* 2017–2027. Thimphu, Bhutan: Higher Education Planning Division, Department of Adult and Higher Education, Ministry of Education.

[2] Dema, T. (2018, October 3). *STEM Olympiad to showcase scientific innovation, The Bhutanese.* Retrieved April 1, 2019 from https://thebhutanese.bt/stem-olympiad-to-showcase-scientific-innovation

[3] Dorji, U. (2014). *iSherig Education ICT Master plan 2014–2018.* Thimphu: Bhutan: Ministry of Education, Royal Government of Bhutan.

[4] Ghalley, K. (2019, May 26th). *Annual Education Conference Ends with 16 Resolutions. Business Bhutan.* Retrieved May 26, 2019 from http://www.businessbhutan.bt/2019/01/08/annual-education-conference-ends-with-16-resolutions

[5] Namgay, K. (2017, October 9). *Gyalpozhing College of Information Technology formally inaugurated. Bhutan Today.* Retrieved May 26, 2019 from http://www.bhutantoday.com.bt/gyalpozhing-college-of-information-technology-formally-inaugurated

[6] Palden, T. (2018, June 26). *Bhutan's population is 735,553, Kuensel.* Retrieved March 30, 2019 from http://www.kuenselonline.com/bhutans-population-is-735553

[7] Royal Governance of Bhutan (2010). *Tertiary Education Policy of the Kingdom of Bhutan.* Thimphu, Bhutan: Tertiary Education Division, Department of Adult and Higher Education, Ministry of Education.

[8] Royal Government of Bhutan (2015). *The Royal University of Bhutan Strategic Plan 2013–2020.* Thimphu, Bhutan: Department of Planning and Resources Office of the Vice Chancellor: Royal University of Bhutan.

[9] Royal Government of Bhutan (2018). *Annual Education Statistics.* Thimphu, Bhutan: Policy and Planning Division, Ministry of Education.

[10] State of Tertiary Education (2018). *Department of Adult and Higher Education, Ministry of Education.* Thimphu, Bhutan: Higher Education Planning Division.

[11] Ura, A. and Zangmo, W. (2012). An Extensive Analysis of GNH Index. Thimphu, Bhutan: The centre of Bhutan studies.

[12] Wangmo, L. (2019, January 10). *Sherig Conference Endorses 11 resolutions. Bhutanese Bhutan.* Retrieved May 26, 2019 from https://www.business bhutan.bt/2018/01/10/sherig-conference-endorses-11-resolutions

Building Scientific Temper in Mauritius

Dayachand Balgobin
Department of Mechanical Engineering, Rajiv Gandhi Science Centre, Old Moka Road, Bell Village 11202, Mauritius
E-mail: balgobin.rgsc@gmail.com

ABSTRACT: This paper provides an overview of the role of the Rajiv Gandhi Science Centre (RGSC) along with its achievements since its opening in 2004. An outline of the various flagship activities and programmes that have gained immense popularity among school students is also highlighted.

After14 years of successful operation, the centre is currently expanding its horizon by implementing another unit (Satellite Centre) of the RGSC with updated infrastructure comprising Planetarium and few modern exhibition galleries. The new satellite centre is also being planned to have a state-of-the-art facility—a 125-seater Planetarium for the first time in Mauritius. Along with the new satellite Centre, the Rajiv Gandhi Science Centre is also presently working on revamping most of its permanent galleries of the existing building that as stated above are more than 15 years old.

The Rajiv Gandhi Science Centre (RGSC) being the only institution in Mauritius set up with the collaboration of the National Council of Science Museums (NCSM), India for the popularization of science among school students and public in general is now definitely the hub of Science and Technology promotion in Mauritius.

The second part of the paper will provide an overview of the new strategic plan recently formulated by the Rajiv Gandhi Science Centre with a view drastically review the Centre organizational structure and major key performance indicators in the next 10 years to come. With the six major areas explored in the strategic plan the Centre is looking at completely changing its passive role into a very dynamic one by focusing on Research and E-promotion of Science as well.

Keywords: Mauritius, Rajiv Gandhi Science Centre, Informal Science Education.

1. INTRODUCTION

Why is science important? Can we afford to have scientific illiterate nation? Do we need scientific literate nation with a view to train adequate technical proficient people to complement its economic and defense needs or do we need scientific literate nation just to have wider citizenship goals? The answers to each of these individual questions can be debated at length, however for the purpose of this paper; focus will be made mainly on the crucial role of science centres/science museums can play in our modern society.

Science centres derive their existence mainly from the mushrooming of Science museums during the period of Industrial Revolution. During that period rich Aristocrats were collecting objects with a view to showcase the achievements of science and technology of that time. Later on the private collection of such objects were displayed such exhibition ultimately gave rise to science museums. One such exhibition was the Great Exhibition in The Crystal Palace back in 1851 which is now known as the London's Science Museum. This movement of science-based exhibition soon reached the shores of America, and among others the Science Museum in Boston opened its door in 1864. At present we have around 275 science museums/ science centres in United States only. Gradually over the years the number of such centres science museums/science centres started going up and today we have reached over 750 worldwide. Over time there was a pressing need to create appropriate networks with a view to provide an appropriate platform for science museums and science centres to join efforts together and share common goals.

The Association of Science-Technology Centres (ASTC)[1]: Founded in 1973, in United States, ASTC that provides opportunities for science centres, museums, and related intuitions to mutually share valuable contributions of its members in the field of informal STEM learning. ASTC now represents approximately 700 members in nearly 50 countries, including not only science centres and museums, but also nature centres, aquariums, planetariums, zoos, botanical gardens, and natural history and children's museums.

European Network of Science Centres and Museums (Ecsite)[2]: Created in 1990, Ecsite has the mission to inspire and empower science centres, museums and all organisations that engage people with science, and to promote their actions. It currently has over 350 members which are mainly science centres, museums, planetariums.

The Popularization Network of Science and Technology of Latin America and the Caribbean, (RedPOP)[3]: Created in 1990, in Rio de Janeiro, RedPOP is an interactive network that brings together groups, programs and science and technology popularization centres (S&T). At present, the RedPOP has approximately 80 members, working with the various Centres for the popularization of science and technology in many countries of the world.

The Asia Pacific Network of Science & Technology Centres (ASPAC)[4]: Initiated in 1997 ASPAC is an association of science centres, museums and related organisations, with a regional focus on the Asia Pacific region. ASPAC has around 80 members from over 20 countries and administrative regions in Asia and the Pacific, Europe, Middle East and North America.

Southern Africa Association of Science and Technology Centres (SAASTEC)[5]: Established in 1902 with 60 members SAASTEC has one of its mission spelt out clearly to contribute to the improvement of life of Southern African nations by improving scientific knowledge and skills through the utilization of interactive living science and technology discovery centres (S&T centres).

The National Council of Science Museums (NCSM)[6]: Established in 1978, in India it primarily works in enhancing public engagement and appreciation of Science and Technology through a wide network of 47 hands on and interactive science museums/centres spread across the country.

The reasons behind the existence of such a large number of science centres/science museums can be many like, showcasing objects related to the technological prowess related to science and technology, reasons can be purely financial with considerable economic returns as well but above all these reasons venues like science centres/science museums do undoubtedly provide a learning platform for the visitors visiting them. To some these venues create a passion and love for scientific subjects and ultimately a choice for a scientific career. It is with this aim that the project of making the first science centre came up in early 1990 and was fully materialize and inaugurated in December 2004 with a view to possibly address the issue of the decline of student opting for science and technological field at school level and/or at undergraduate level.

2. THE RAJIV GANDHI SCIENCE CENTRE

Rajiv Gandhi Science Centre (RGSC) was set up based on Memorandum of Understanding between the then Ministry of Education & Human Resource Development, Mauritius, and the National Council of Science Museums of India on 3 October 1998. As per the Memorandum of Understanding, Government of India collaborated by providing expertise in designing of the exhibition galleries, supplied all the exhibits and teaching aids and the Mauritius Government's main role was to provide the land, construct the building and run the centre thereafter.

In view the vast experience India in running science centres the National Council of Science Museums (NSCM), Kolkata was mandated to design and fabricate exhibits for the exhibition galleries, the outdoor science park, a Mobile Science Exhibition for RGSC. NCSM had been closely involved in the planning of the Science Centre interacting on technical matters relating to the construction of the building. Government of Mauritius had provided land, funds for construction and given grant to Rajiv Gandhi Science Centre Trust Fund for running of the Centre.

The RGSC was inaugurated on 30 November 2004 by Mrs. Sonia Gandhi and since then, the Centre has received more than 400,000[7] visitors. Operating presently under the aegis of the Ministry of Education, Human Resources, Tertiary Education, Science, Research and Technology, RGSC has one of its main objectives, to inculcate scientific awareness in current and relevant fields and arouse curiosity among the public in general. After 20 years of successful operation, RGSC has realized its vision of becoming a Centre of excellence in the communication and promotion of Science and Technology (S&T).

2.1 Promoting Scientific Awareness and Scientific Literacy

Curiosity is an innate quality of many species including humans. That goes without saying therefore that one can hardly dissociate curiosity from human beings. Venues like science centre/science museums encourage curiosity with beautifully designed exhibits that are in many cases nice, appealing, surprising or even funny. In the words of Frank Oppenheimer, founder of San Francisco's Exploratorium, "No one ever failed a museum."[8] For some, the interests awakened by science centre experiences have turned into a passion for science, and the beginning of a lifetime devoted to teaching or research.

Quite often many people in the field of science popularization talk about scientific awareness and scientific literacy without much demarcation between the two. Some argue that mass media and TV programs[9] covering issues related to global warming, other environmental issues, and health pandemics do contribute in one way or the other towards scientific awareness. This can be considered as a casual manner of being exposed to issues related to science and technology.

On the contrary, scientific literacy is a little more structured way of getting involved with science related issues. Formal teaching of science allows one to understand the world around them; this indeed contributes towards science literacy. Scientific literacy as defined by Leonie J., Rennie (2005)[10] talks about making informed decisions about the environment and one's own health, also talks about interest in and understanding of the world around us.

With the above perspective as background venues like science centres, science museums, its more about creating science awareness although it might as well be contributing toward science literacy. The Rajiv Gandhi Science Centre has 5 permanent galleries out of which 3 talks about the origin of Mauritius as a volcanic island although the major part is about the evolution of the Universe, the second one talks about the land and environment of Mauritius and the third one deals mainly with the Resources of Mauritius. These three galleries give a scientific insight regarding Mauritius as a volcanic island giving a different picture of the island as compared to its historic representation often mentioned media and other existing publications.

The two remaining galleries allow one to explore the feats and achievements in the field of science and technology and learn basic principle of science in a very playful manner. School students, families and tourists spend as long as half a day visiting the various attractions.

The Rajiv Gandhi Science Centre through its various activities do promote scientific literacy to a great extent. It is worth mentioning two activities that has gained much popularity among the school students. One is the *Science Quest* and the second one is the *Model Glider Competition. Science Quest* provides the school students the opportunity to investigate issues related to science and technology using the scientific method. Here science literacy forcefully comes into play in such circumstances. The student no doubt engages in some sort of discourse with scientific issues, they are able to identify, question, investigate and draw evidence-based conclusions. The other highly demanded activity is the *Model Glider Competition* which also fosters scientific literacy among school students. After being initiated with basic principles of aerodynamics through professionals in the field, the students gets the chance to build their own prototype of model gliders, fly them and thereafter refine their prototype to improve the flying distance.

2.2 Centre of RGSC—The Planetarium

While planning, deciding and executing STEM promotion which is normally associated with the promotion of basic sciences like chemistry, physics, biology and mathematics we often do not realise that above all this, the promotion of Astronomy is an important experience that has far-reaching implications for many curriculum areas. It is strongly felt that "Astronomy literacy"[11] should definitely be part and parcel of science literacy.

Whether one is literate, or simply illiterate, or whether one is science literate or totally ignorant about scientific issues, the sky has been drawing attention of one and all since pre-historic times people at large have been fascinated with what is up there in the sky.

The Rajiv Gandhi Science Centre has not been ignoring this area of science and though various means like the night sky watch programme it has been touring the island since nearly two decades now. The most appreciated are the programme around celestial events, like transit of Venus that was visible from Mauritius and Solar eclipses and sky observation programmes in various location of the country. The recent solar eclipse in September 2016 attracted around 5,000 visitors to the centre in a single day and around 17,000 special solar eclipse glasses were also sold to school students and public.

Over the years it is because of the zeal of the general public of as regards to the activities related to celestial phenomenon and also because of the very high demand of the sky observation programmes that gave rise to the conceptualization of the upcoming planetarium of the Centre. For a more enriching and more immersive experience, a 12 m dome planetarium is being planned and the process for execution of the project is on

With regards to the choice of the projection system for the Planetarium of the RGSC we had the choice between two systems widely in demand currently. We could have either opted for a Hybrid projection system of a fully digital projection system. The Hybrid planetarium systems combines the opto-mechanical projectors for best viewing of starry objects along with fulldome digital video system allowing colourful images to the dome simulating movement of stars and spaceflights for example. On the other hand, the fully digital dome projecttion system has a numerous advantage including that of providing a fully immersive experience with more flexibility. The Centre has however been opting for a digital projection system in view of the various advantages of such asystem.

2.3 Strategic Plan of RGSC—Focusing on 6 Major Areas[12]

Following the planning of the Rajiv Gandhi Science Centre since 1998 and the final set up, tremendous efforts have been made to reach out all segments of the target group with various programmes to raise awareness in the field of science and technology and supplementing school education as well.

However, there was a strong need to have an overarching plan of strategic moves for the coming decade or so. It is with that perspective in mind that the very first strategic plan (2017–2030) was made and launched in 2017.

The plan focusses on six major areas, namely the Centre itself, Visitor management, Supplementing Education System, Outreach, E-Promotion of Science and lastly Research and Development.

The plan defines various strategic objectives and action plan along with key performance indicators with a view to remarkably improve the main infrastructure, services and deliverables of the centre as a whole. New targets have been set to provide a more pleasant and functional environment at the venue, procure on a regular basis high value exhibits, new attractions thus ensuring repeat visits. These are snapshots of some of the strategic targets set up under the first focus area which is the Centre itself and its services.

Visitor management is the second focus area which details elaborate objectives to have customer focus approach along with improved quality of service. This area also talks about the urgency to have appropriate marketing plan and innovative programmes that could guarantee constant influx of visitors all throughout the year.

The third focus area is supplementing school education system considering school students as the Centre's main target group. This area although having been well taken care since nearly two decades however still requires some utmost attention because it forms part is one of the very crucial mandate of the centre. The strategic plan enumerates measures to develop and adopt an inclusive approach in the dissemination of science and technology paying particular attention to school requiring special needs.

The fourth focus area is outreach—bringing science to the door steps of each and every Mauritian. The centre organizes lots of outreach programme with a view to reach out the population segment that for one reason or the other fail to make visits to the centre. Strategies have been defined to increase networking with stakeholders and social partners to organize more community-based activities and events.

E-promotion of science is the fifth area that the strategic plan has delved upon. RGSC has not really been tapping the full potential of E-promotion of science. With the new advancements made in the field of information and communication technology and increased accessibility of internet services, it was high time therefore for RGSC to define ways and means to maximize visibility and output by tapping all online available resources and embark on the development of android applications for forthcoming activities and events.

The last area of interest of the strategic plan is Research and development which though very important but for one reason or the other till date was not given much attention. We all know how research and development plays a very important role in the success of any organization. The first two major task in this area is to firstly assess the status of science and technology in the island and secondly based on the previous study strengthen the sustainability of STEM promotion by RGSC.

3. CONCLUSION

After nearly 20 years of operation although it is too early to say "mission accomplished" but RGSC has been coming a long way. The first strategic plan (2017–2030) is a very well elaborated roadmap towards a definitely more impactful achievement of the centre. Various key actions as defined in the strategic plan is already being implemented with really encouraging results but we still have a long way to go. To end there is no doubt that the Rajiv Gandhi Science Centre is the only location or we can say a one stop shop for science awareness, promotion and communication activities definitely contributing towards building scientific temper in Mauritius.

ACKNOWLEDGEMENT

I wish to thank the Director to the Rajiv Gandhi Science Centre for his kind support and cooperation. He has been instrumental in helping the author in framing the paper and also readily allowed the reproduction of small extract from the RGSC strategic plan 2017–2030 to be included in the present paper.

REFERENCES

[1] Home–Association of Science and Technology Centers. Retrieved from https://www.astc.org
[2] Ecsite—The European Network of Science Centres and Museums. Retrieved from https://www.ecsite.eu
[3] Red de Popularización de la Ciencia y la Tecnologíaen América latina y el Caribe (RedPOP). Retrieved from https://www.redpop.org/red-de-popularizacion-de-la-ciencia-y-la-tecnologia-en-america-latina-y-el-caribe-redpop
[4] The Aspacnet.org. Retrieved from https://www.aspacnet.org/history
[5] SAASTEC—Southern African Association of Science and Technology Centres. Retrieved from https://saastec.co.za
[6] Home—National Council of Science Museums. Retrieved from https://ncsm.gov.in
[7] Rajiv Gandhi Science Centre Annual Report 2017/2018.
[8] Cynthia Yao *et al.* (2006). Handbook for Small Science Centre. Page 247.
[9] Richard (2009). Public Awareness and General Understanding of Science and Technology.

[10] Rennie, Leonie (2005). Science Awareness and Scientific Literacy. Teaching Science 51(1), 10–14.

[11] Great Lakes Planetarium Association. Astronomy Literacy: Essential Concepts for a K-12 Curriculum. Retrieved from https://glpa.org/k12concepts.

[12] Rajiv Gandhi Science Centre. Strategic Plan 2017–2030.

Science, Technology and Innovation in Togo

Yao Adjrah*, Sename Dodzi Kossi and Kouami Kokou
Directorate for Technical Scientific Research (DRST), Lome, Togo
E-mail: *directiondelarecherche_togo@yahoo.fr

ABSTRACT: Togo has committed itself to the use of Science, Technology and Innovation (STI) as defined through the various National Development Policy Documents. Togo possesses a National Policy in STI. The goal of this policy is to provide the nation with a system of research, innovation with the aim of improving the living conditions of the populations and contribute to the process of emergence of the country by 2030.

STI policy is not addressed exclusively to governments, scientists and companies. It is focused at all components of Togolese society. In this respect, it pursues three ultimategoals: (i) to contribute to the well-being of Togolese society; (ii) contribute to common prosperity in a perspective of sustainable, harmonious and equitable development throughout the territory; and (iii) enrich the Togolese culture and contribute to the world heritage of knowledge.

The population of Togo is highly active (70%) but with little employment. The country has a certain potential in terms of agricultural production: cocoa, coffee, maize, fiber, cottonseed. On the mining plan, the Togolese subsoil is full of phosphate, clinker, iron and a large reserve of ornamental stones. Togo has a deep-water port. The exploitation of these agricultural and mining potentials provides little added value because of the low rate of transformation in the country.

The main areas targeted by STI and the subject of research are: 1) natural sciences; 2) biomedical sciences, pharmaceuticals and traditional pharmacopoeia; 3) agricultural sciences, nutrition and food; 4) environmental sciences; 5) human and social sciences; 6) legal, economic and management sciences; 7) industry and crafts; 8) housing and urban planning.

The country has limited research capacity in relation to infrastructure (buildings, equipment and consumables in all fields). In general, human resources are also limited and not very diversified. Most researchers are lecturers-researchers. The number of PhD degree researchers who devote almost all their time to research is very low. The representation of women is negligible in national research institutions. The number of technicians and other support personnel is negligible. National research institutions obtain results that are mainly publication-oriented and almost non patent-oriented. The level of technology transfer from laboratories to production sectors is also low despite the availability of results in institutions. Funding for research not specified but is embedded in the overall budget allocated to institutions that perform other activities in addition to Research-Development and Innovation activities.

Many constraints limit the promotion of STI in Togo: scientific culture, brain drain, the legislative and political framework, equipment, financing, human resources, etc. But the main legislative constraints to STIs today remain the absence of a National Research Policy and the non-adoption of the National STI Policy validated technically in 2014. The absence of these guidance standards does not facilitate the development of scientific and technological research activities. There is a low availability of heavy analysis equipment, or advanced analyses in laboratories for research, but also difficulties in renewing obsolete equipment and maintaining existing equipment. The budget devoted to research is insufficient.

The perspectives are: (1) adoption of the Togo STI Policy; (2) finalization and adoption of the National Research Policy; (3) Establishment of a competitive fund for scientific research and innovation.

Keywords: Togo, National Research Policy, Technology, Innovation Policy, Private Sector.

1. INTRODUCTION

Science and technology are perceived the world over as major tools for rapid social and economic development. The more industrialized countries of the world applied science and technology to develop their economies (Ministry of environment, science and technology, 2009). Togo (West Africa) aspires to become an emerging nation by 2030. Achieving this ambitious vision will necessarily involve, among other things, a more effective application of science, technology and innovation (STI) in economy in order to drive productivity growth and diversification in production (SCAPE, 2013). Efforts have been made to set up the appropriate regulation, but the implementation and functioning of the structures for the real promotion of science, technology and innovation (STI) is insufficient. The role played by STI in the country's development has been limited. This report describes the STI situation in Togo.

2. STATE OF STI SYSTEM IN TOGO

2.1 Legal and Regulatory Framework

2.1.1 *Regulatory Framework of Research in Togo*

The research sector in Togo falls under the governance of the Ministry of Higher Education and Research (Decree N° 2005-069/PR of 05 August 2005). The Orientation Law of Higher Education and Research places special emphasis on the development of research:

- Coordination of all structures and research activities;
- Capacity building for dissemination, popularization, consultation, reproduction and management of scientific and technical knowledge;
- Research results valorization and the protection of the intellectual property;
- Strengthening the public-private partnership in the artistic, scientific, technological and innovation fields;
- Technology transfer and capacity building.

The higher education focus is to organize research and technological innovation in order to promote the potential and available natural resources of the country.

2.1.2 *Main Research Governance Bodies*

- The national research council is responsible for defining national research guidelines. It studies and appreciates the results of the research activities carried out on the national territory.
- The inter-ministerial technical committee for research gives its opinion on any research activity conducted on the national territory.
- Directorate of Scientific and Technical Research ensures the coordination and control of the research institutions, under the authority of the minister responsible for research.

2.1.3 *Regulatory Framework of STI (ethics, intellectual property)*

The regulatory framework of STI in Togo takes into account a number of provisions relating to ethical issues and management of intellectual property.

2.1.3.1 *National Bioethics Advisory Committee (NBAC)*

The NBAC is a multi-disciplinary and multi-sectoral body with the objectives: (i) conduct ethical reflections on issues in the field of bioethics; (ii) play an advisory role through the formulation of guiding principles and definition of conduct rules for biomedical research; (iii) promote education and awareness of bioethics.

2.1.3.2 *Bioethics Committee for Health Research (BCHR)*

BCHR's main task is to review research proposals and supporting documents, with particular attention to the process of obtaining informed consent, documentation, and the relevance and feasibility of the protocol.

2.1.3.3 *Intellectual Property in Togo*

The administration of intellectual property rights at the national level comprises three structures:

- National Institute of Industrial Property and Technology (NIIPT) dealing with industrial property issues. It is responsible among others to:
    - Disseminate all relevant information and technical documentation relating to industrial property;
    - Sensitize potential users (researchers, inventors, innovators, industrialists, craftsmen, traders, farmers, etc.) on the essential notions of industrial property and on the interest of the documentation constituted by patents for research technical and development;
    - Assists economic operators: (i) for the signing of agreements for the sale and sale of licenses, know-how, technical assistance and training of personnel; (ii) for the choice of technology equipment.

- Togolese Copyright Office ("Togolese Copyright Office," BUTODRA) dealing with copyright issues;
- National Council of Intellectual Property (NCIP), an advisory body that strengthens the actions of NIIPT and BUTODRA. It advises on all matters relating to draft conventions or treaties related to intellectual property.

2.2 Institutional Frame

2.2.1 *Main Bodies Responsible for STI*

The institutional infrastructure of research in Togo includes:

- Public universities;
- Public research institutes;
- Research and development units in companies;
- Branches of international research institutions.

The most important research centers in Togo:

- Lomé University with its laboratories and research centers;
- Kara University with its laboratories;
- Togolese Institute of Agronomic Research (ITRA);
- Research structures in Pharmacopoeia;
- Building and Housing Center (CCL);
- International Center for Soil Fertility (IFDC);
- International Linguistic Society (SIL);
- West African Science Centre on Climate Change and Adapted Land Use (WASCAL)
- National Institute for Scientific Research (INRS);
- Center for Coastal and Environmental Management;
- National Laboratory of Buildings and Public Works (LNBTP);
- Wycliffe;
- Demographic Research Unit;
- National Institute of Statistics and Economic and Demographic Studies.

2.2.2 *Role of the Private Sector*

The involvement of private sector STI activities is very weak. The general framework for innovation is not sufficiently developed in the country and is not suitable for supporting a dynamic innovation ecosystem.

However, the private sector benefits from services and technologies generated by the research structures are not negligible. The services include laboratory analysis, products, etc. But there are rarely any formal contracts with private companies.

2.2.3 *Academies, Scientific Associations, NGOs and Civil Society*

Scientific and technological life in Togo is also animated by NGOs, scientific associations, learned societies, etc.

2.2.3.1 *Academies*

- National Academy of Arts, Science and Technology (ANASTI): has for essential tasks, to study in their various aspects (artistic, scientific, political, economic, technical, historical, ethnographic, biological, social and cultural) the questions the country, and to submit to the competent authorities its opinions, analyzes and suggestions;
- Ewe Language Academy: responsible for conducting research on the main local language spoken in Togo;
- Kabyè Language Academy: responsible for conducting research on the main local language spoken in Togo;
- International Language Society (SIL-Togo): international institution of research on local languages, their writing and popularization;
- Wycliffe: institution of research on local languages, their writing and popularization.

2.2.3.2 *Associations – NGOs – Civil Society Organizations*

In addition to traditional research and STI institutions, many civil society organizations (Associations and NGOs) are also involved in scientific and technological innovation activities. Some of them are listed in Table 1.

Table 2 summarizes the number of institutions involved in the STI system, while Table 3 specifies the knowledge and technologies developed in some of these research institutions.

Table 1: Associations and NGOs Involved in STI

Institution Name	*Denomination*	*Research Fields*
Believe	Center for Action Research for the Environment and Integrated Development	Agricultural sciences
SICOR-SA	Society of Innovations and Rural Construction	Agricultural sciences
CERA	Cabinet of Expertise in Research and Action	Human sciences and society
CPL	Phytopharmaceutical between LAKAZA	Health Sciences
ORIDA, New Year	Scientific Research Centre for Herbal Medicines recipes	Health Sciences
CROP	Centre for Research on public Opinion	Human sciences and society
Grade	Action Research Group for Development Education	Legal Sciences, Politics and Administration
GARDEGI	Research Action Group for Integrated Global Development	Agricultural sciences
COOP-CA-PS-OTISSAN	Agricultural Cooperation Cooperative OTISSAN	Agricultural sciences
BIOPRODUCT		Agricultural sciences
Feel	Sino - Togolese company	Agricultural sciences
CFRET	Tami Rural Training Centre	Agricultural sciences
AVSF	Agronomists and Veterinarians without Borders	Agricultural sciences
USEFUL CODE		Agricultural sciences
CD2A/AREJ	Agroecological and craft development center	Agricultural sciences
BONITA HOUSE TECHNICAL UNIVERSITY		Agricultural sciences
CARTO	Rural Animation Center of Tambimong Ogaro	Science and Technology
Raffia	Research Support and Training in Self-Development Initiatives	Science and Technology
APEEERTO	Association of Promoters of Energy Efficiency and Renewable Energies	Science and Technology

Table contd...

Table 1 contd...

Institution Name	*Denomination*	*Research Fields*
OPEDRA	Observatory of Population and Issues for the Development of Africa	Science and Technology
AVOVIKA	Association of Volunteers for Garbage Collection in the City of Kara	Environment
AJAC	Association of Youth for Sanitation and Construction	Environment
ETD	Businesses, Territories, Development	Environment
NSCT	New Cotton Company of Togo	Agroecology
ANPAT	Avian sector professionals association of Togo	Avian sciences
MAPTO	Movement for a Peasant Alliance in Togo	Agricultural sciences
FNGPC	National Federation of Togo Cotton Producer Groups	Agricultural sciences
FUPROCAT	Federation of Coffee-Cocoa Producer Unions of Togo	Agricultural sciences
CTOP	Togolese Coordination of Farmer Organizations and Agricultural Producers	Agricultural sciences
CPC	Togo Center of Cereal Producers	Agricultural sciences
CIDAP Baga	International Center for Agro Pastoral Development	Agricultural sciences
LASEE	Laboratory for Sanitation in Water and Environmental Sciences	Science and Technology
LaBAN	Laboratory of Food Biochemistry and Nutrition	Science and Technology
Prades	Pole of Research and Expertise on the Dynamics of Spaces and Societies	Science and Technology
LaREA	Research Laboratory in Applied Economics	Science and Technology

Source: MESR, 2018.

Table 2: Number of Institutions Involved in the STI System

	Public	*Private*	*Total*
Universities	02	03	05
Research Centers	10	02	12
Research Institutes	02	–	02
Research Laboratories	32	–	32
NGO – Associations	–	35	35
Total	46	40	51

Table 3: Examples of Knowledge and Technologies Developed in Research Institutions

Research Institutions	*Knowledge/Technologies Developed*	*Targeted Production Sectors*
Public universities	• Processing and conservation of agriculture resources (fruits and vegetables, cereals, tubers, spices) • Protection against pests; • Herbal recipes for the main diseases (Malaria, diarrhea); • Treatment of urban waste in compost; • Exploitation of solar energy; • Production of day-old chicks and improved poultry feed formulation processes • Protection against coastal erosion and weather forecasting; • Valorization of plant species for timber.	- Nutrition and food - Agriculture - Health - Environment - Energy - Building
Togolese Institute for Agricultural Research	• Varietal and breed improvement • Improving management of agricultural production systems • Improvement of soil fertility • Food Technology	- Agriculture - Environment
Construction and Housing Centre	• Cooked brick • Dolomitic lime	- Building - Housing
IFDC	• Improvement of soil fertility	- Agriculture - Environment
SIL	• Development of local languages • Alphabet design	- Culture

2.3 Incentive Measures for Research and Innovation

Some incentives for research and innovation have been identified in Togo.

2.3.1 *Information and Access to the Internet*

2.3.1.1 *Internet Coverage of STI Research Infrastructures*

Although the situation is far from ideal, some recent initiatives have made it possible to strengthen the Internet coverage of STI infra structure. The biggest initiative is the E-Gov network whose work was launched in June 2014, and which provides the administration with the necessary tools to modernize. All the buildings of the ministries, the institutions of the Republic and the public structures of STI (universities, centers, institutes, etc.) are connected.

The second initiative is called "Wifi Campus". Through this project, public universities and University Hospital Centers of Togo now have broadband internet accessible to all.

2.3.1.2 *Access to Scientific Information (physical and online libraries)*

Through INASP (International Network for Availability of Scientific Publications), faculty members and researchers have access to online journals and journals such as:

1. American Institute of Physics journals
2. Annual Reviews
3. British Institute of Radiology Journals
4. Geological Society—The Lyell Collection Complete
5. IMF eLibrary (Your online access will expire on May 8th, 2020)
6. JSTOR
7. Liebert Online
8. Liverpool University Press
9. OECD iLibrary
10. OUP-E-Books American National Biography
11. OUP-E-books Dictionary of National Biography
12. OUP-E-books Grove Art Online
13. OUP-E-books Grove Music Online
14. OUP-E-Books Oxford English Dictionary Online

15. OUP-E-books Oxford Reference Online
16. OUP E-books Oxford Scholarship Online
17. OUP Oxford Journals
18. Oxford Biblical Studies Online
19. SPIE Digital Library.

2.3.2 *Incentive Allowances for Research and Publications in Indexed Specialty Journals*

In the two public universities where most of the STI activities are carried out, research and library allowances are incorporated into the salary of the lecturer-researchers.

2.4 National System for Dissemination of Research Results

STI researchers have a variety of publishing institutions and newspapers to publish their work as listed in Table 4. At the university level, scientific days are organized every two years but attract very few private sector entrepreneurs and the general public.

Table 4: List of STI Publishing Institution and Scientific Journals

No.	*Name of the Publication Institution*	*Type*	*Publication Form*	*Faculty/Research Structure*	*ISSN/ISBN Number*	*Periodicity/ Publication Frequency*
1.	Togolese Studies: Togolese Journal of Science	Journal	Paper	National Institute for Scientific Research (INRS)	ISSN 0531-2051	2 times per year
2.	Togolese Science and Innovations (FTSTI)	Technical sheet	Paper	National Institute for Scientific Research (INRS)	978-2-37745-000-8/EAN	Variable
3.	Scientific, human and social notes (NSHS)	Journal	Paper	Faculty of Humanities and Social Sciences	2409-9791	2 times per year
4.	Mr. Nunya. Philosophy, scientific and technical heritage	Journal	Paper	Faculty of Humanities and Social Sciences (FSHS)/Laboratory HIPHIST	2311-3375	1 time per year

Table contd...

Table 4 contd...

No.	*Name of the Publication Institution*	*Type*	*Publication Form*	*Faculty/Research Structure*	*ISSN/ISBN Number*	*Periodicity/ Publication Frequency*
5.	Transmit	Journal	Paper	National Institute of Education Sciences	2414-8156	2 times per year
6.	Multi Fontaines	Journal	Paper	Faculty of Literature, Languages and Arts	2311-1658	1 time per year
7.	Environmental Science Review (Rev. Sc. Env.)	Journal	Paper	Biogeographical Research and Environmental Studies Laboratory (LaRBE)	1812-1403	1 time per year
8.	Action	Journal	Paper	Faculty of Literature, Languages and Arts/ Research Center for Black Africa and Diaspora Literature CReLAND	2071-1964	2 times per year
9.	Exchanges	Journal	Paper	Laboratory of Analysis of Political, Legal, Economic and Social Changes	2310-3329	2 times per year
10.	Literatures and civilizations	Journal	Paper	Faculty of Literature, Languages and Arts	2410-2105	2 times per year
11.	AHOHO	Journal	Paper	Faculty of Humanities and Social Sciences/ Dynamics of Environments and Societies research Laboratory	1993-3134	2 times per year
12.	Trifop, LETRIA review	Journal	Paper	Faculty of Humanities and Social Sciences/LETRIA	2409-8000	1 time per year
13.	Journal of Humanities and Humanities, ANYASA	Journal	Paper	Faculty of Humanities and Social Sciences	2312-7031	2 times per year

In most cases, national research institutes obtain results that are mainly publications-oriented and not patent-focused. The level of technology transfer from laboratories to production companies is also low, despite the availability of results in institutions.

The national system of dissemination of results is poorly organized. The few journals and scientific journals have irregular issues.

2.5 Infrastructure/Equipment and Funding

The country has limited capacity in terms of infrastructure and research equipment (buildings, measuring equipment and consumables in all areas). As for the existing, their obsolescence and the lack of maintenance makes them not very operational and exposes users to permanent risks of insecurity.

The funding of research is provided by the Togolese State, the international development partners of Togo, and bilateral or multilateral cooperation (AUF, OIF, IDRC, GTZ, UNDP, WHO, FAO, IFAD, UNESCO, French Cooperation, Cooperation and Cultural Action Service, etc.) and Foundations, Academies and some NGOs (NAM'S Center, TWAS etc.).

Funding for research by the state is not specified but is embedded in the overall budget allocated to institutions that perform other activities in addition to research-development and innovation activities. This funding is mainly used to pay the salaries of research staff.

2.6 Human Resources

Most researchers in Togo are employed by the public sector (89.65%) (Table 5). PhD level researchers are dominant (76.77%) (Table 6).

Table 5: Number of Researchers by Type of Institution

Type of Institution	*Number of Researchers*	*Percentage*
Public	710	89.65
Parapublic	8	1.01
Company	7	0.88
Free/without institution	16	2.02
NGOs	51	6.44
Total	**792**	**100.00**

Source: DRST&INRS, 2017.

Table 6: Researchers According to their Educational Level

Educational Level	*Number*	*Percentage*
"BAC"	6	0.76
Technician	24	3.03
Bachelor	5	0.63
MS/MSc, MA ("Master")")	19	2.40
Master/(NVQ Level 5) - ("DEA"/ "DESS")	74	9.34
Master's Degree in Engineering	54	6.82
Doctorate/PhD	608	76.77
"DES"	2	0.25
Total	**792**	**100.0**

Source: DRST&INRS, 2017.

Researchers from Togo (a total of 792) are divided into several fields of research: Agronomic Sciences; Science and Technology, Humanities and Social Sciences, Health Sciences, Language and Literature Sciences, Legal and Economic Sciences, Industry and Handicrafts (Table 7). Further, a low representation of women in the research sector is recorded in Table 8.

Table 7: Institutions by Research Field

Research Field	*Number of Research Institutions*	*Percentage*
Handicraft	5	5.32
Literature and Languages	7	7.45
Multidisciplinary	3	3.19
Agricultural sciences	46	40.94
Human sciences and society	14	14.89
Health Sciences	7	7.45
Science and Technology	10	10.64
Economics, Law, Politics and Administration	2	2.13
Total	**94**	**100.00**

Source: DRST-INRS, 2017.

Table 8: Number of Researchers by Sex

Sex	*Number*	*Percentage*
Male	743	93.93
Female	49	6.07
Total	792	**100.00**

Source: DRST&INRS, 2017.

3. ANALYSIS OF THE CONSTRAINTS

Several constraints limit the promotion of STI in Togo: 1) the legislative and political framework; 2) brain drain; 3) infrastructure/equipment; 4) financing; 5) human resources.

3.1 Legislation and Political Framework

"The law establishing" the Higher Education and Research orientation states clearly that "Every institution of higher education and research must adopt an innovation program, and periodically submit a special report on the innovations introduced to improve (Togolese Republic, 2017) the economic growth and the quality of life of citizen." The mechanisms for applying this orientation law were not implemented. In addition, the small funds allocated to research and innovation do not facilitate the implementation of existing legislation.

But the main legislative constraint to STIs today remain the absence of a National Research Policy. Furthermore, the National STI Policy validated technically in 2014 is not yet adopted by parliamentary committee. The lack of these guidance tools does not facilitate the coordination and the development of scientific and technological research activities.

3.2 Brain Drain Phenomenon

Literature review shows that Togolese migration to OECD countries is relatively skilled. According to the second edition of the World Bank's Migration and Remittances Statistics 2011 (ICMPD and OIM, 2015), emigration affects up to 18.7 per cent of Togo's tertiary graduates. Many become great researchers, doctors, highly qualified staff, university professors, teachers, engineers but choose to stay in the host countries rather than return to the country of origin (Togotopnews.com, 2019). The presence of these experts abroad indicates that the country is facing a brain drain phenomenon reducing the capability to boost the innovation sector.

3.3 Infrastructure and Equipment

The country has limited research capacity through research infrastructures (buildings, measuring equipment and consumables in all fields). Their dilapidated infrastructure and equipment make them less operational and exposes users to permanent risks of insecurity. Also, there is a low availability of heavy analysis equipment, or advanced analyzes in laboratories for research, but also difficulties in renewing outdated equipment and maintenance of existing equipment.

3.4 Funding

The system for allocating funds for research is not subjected to the competitive pressure which drives relevant and novel research tailored to the needs of the nation.

The STI system in Togo records a lack of material and financial means. Indeed, the funding granted by the government to research institutions represents less than 0.38% of GDP, whereas the recommendations of the Lagos Plan of Action and the document of Politics in Science, Technology and Innovation of ECOWAS, adopted by the Heads of State and Government, indicate an amount of at least 1% of GDP.

3.5 Human Resources

Human resources are generally limited and undiversified. Most researchers are lecturer-researchers. The number of doctor-level researchers who devote almost all of their time to research is very small. This situation is partly linked to the lack of real status of researchers and especially the lack of motivation that discourages research orientation in a specific way.

4. KEY ELEMENTS OF STI POLICY NOT YET ADOPTED

4.1 Main Objective

Define a global framework to use Science, Technology and Innovation in the implementation of the Accelerated Growth and Employment

Promotion Strategy to achieve socio-economic transformation and the emergence of Togo.

4.2 Specific Objectives

The specific objectives of this TSI Policy are to:

- Allow Togo to take its place in the knowledge society;
- Promote Science, Technology and Innovation to sustainably accelerate economic growth and ensure social well-being;
- Use Science, Technology and Innovation to contribute to job creation in all priority sectors of the Togolese economy;
- Promote a scientific culture with real impacts on the environment and the management of the country's natural resources.

4.3 Strategies for Achieving the Stated Objectives

- Improving Governance of Science, Technology and Innovation
- Capacity building in human resources
- Strengthening the capacity of production, dissemination, extension and management of scientific and technical knowledge
- Implementing research and innovation infrastructures
- Promoting Public-Private Partnership in Science, Technology and Innovation
- Protecting Intellectual property
- Promoting effective management of data on STI performance indicators.

4.4 Priority Domains

The priority domains are: Education, Agriculture, Health, Energy, Water and Environment, Mines, Transport, Information and Communication Technology, Industry, Fauna, Flora and Tourism, Defense and Homeland Security, Shipping and Related Activities, Trade, Building and Public Works, Social Sciences and Humanities,

5. CONCLUSION

The work of research institutions in Togo is mainly focused on agriculture. As forces for the research system in Togo, we can note: (1) existence of a research coordination body; (2) existence of a diversity of structures making research; (3) availability of researchers. In terms of weaknesses: (1) under-equipment of research structures; (2) low funding for research; (3) very little certified and patented research and innovation.

For repositioning STI system as development tools, Togo will have to address nine critical challenges:

- Enacting the STI policy document;
- Giving to STI a central role in national economic planning and development;
- Improving governance of Science, Technology and Innovation;
- Training of a critical mass of STI experts;
- Aligning the research and development (R&D) system to socio-economic needs;
- Strengthening the capacity for production, dissemination, extension and management of scientific and technical knowledge;
- Setting up appropriate STI infrastructures/equipment;
- Promoting public-private partnership for research and development;
- Implementating science and technology parks;
- Accelerating the allocation of a minimum of 1% of the Gross Domestic Product (GDP) to support the science and technology sector.

REFERENCES

[1] DRST & INRS (2017). Final report of the study on the research map in Togo, Lome, Togo.

[2] ICMPD and OIM (2015). Survey on migration policies in West Africa, Dakar, Senegal, pp. 317–328.

[3] MESR (2018). Country report 2017 on research and innovation in Togo, Lome, Togo.

[4] Ministry of environment, science and technology (2009). National science, technology and innovation policy, Accra, Ghana.

[5] Republic of Togo (2017). Law n° 2017-005 for higher education and research orientation (of June 19, 2017), Lome, Togo.

[6] SCAPE (2013). Accelerated Growth and Employment Promotion Strategy, Lome, Togo.

SECTION 3

STI Popularization Tools

Use of Visualization as a Tool for Popularization of Science in General Education in Sri Lanka

Madawala Liyanage Shanaka Piyatissa[1*], Mdgaparmdjohar[2] and Arun Kumar Tarofder[2]

[1]Department of Science, National Institute of Education, Maharagama, Sri Lanka

[2]Management and Science University, Persiaran Olahraga, Shah Alam, Selangor, Malaysia

E-mail: *shanakapiyatissa77@gmail.com

ABSTRACT: There involves an overt component of visualization in science and helping students to visualize the needed areas in science using suitable external representations is a way to strengthen students' familiarity with the visualization aspect embedded therein. This study suggests that paying deliberate attention to elaborating the visualization aspect in teaching science in the classroom can have a positive impact on students' science performance which will help popularize science. To prove this, a field experiment was conducted with a group of secondary school students (184 grade 10 students) in Sri Lanka. Selected physics lessons in their school science syllabus rich in visualization content were chosen as the experimental lessons. To assess the students' familiarity with the visualization aspect in these lessons, a lesson and content-specific assessment tool called Visualization Related Physics Test (VRPT) was utilized during the study on which the success of the experiment depends. Identification of visualization related areas in the physics lessons is a major factor of this study. The VRPT contains 28 multiple choice type physics questions capable of assessing the students' attainment in visualization in physics. The VRPT is a reliable and validated tool. To highlight visualization related areas, several computers aided learning (CAL) materials were used with the experimental group supplementing the usual classroom teaching in this study. The VRPT scores showed a significant gain in the experimental group (16.41%) over the control group as a result of the teaching intervention. Similarly, physics marks in school term tests also showed a significant gain of 10.25%. It is proposed here that the application of the visualization in appropriate areas in science lessons would result in improving the secondary school students' science performance.

Keywords: Visualization, Representations, External Representations, Internal Representations, Model, Science Performance, Popularizing Science.

1. INTRODUCTION

In science education, visualization can be taken as any type of representation used to make an abstract scientific concept visible. Representations are two-fold: external representations and internal representations. Diagrams, physical setups, textual and verbal expressions, standard notations and equations, video clips, computer animations and simulations, and role-plays can be mentioned as examples of external representations presently used in classroom teaching. With relevance to the external representations, students build internal representations. These internal representations are called visualizations, mental images or visual-perceptions (Gilbert J., 2010). Also, the visualizations include the temporal/spatial relationships of external representational items which can be static and/or dynamic. Scientific models can be considered as another construct in visualization which is closely related to the representations (Shu-Nu & Yao, 2014). A scientific model is a simplified representation of a phenomenon, which is used to simulate its functionality (Piyatissa, Jahar & Tarofder, 2018).

Although visualization is a complex process which includes cognitive and brain activities (Reiner, 2008), effective use of the external representations to elaborate the visualization aspect of science can be considered as a practical way to help students to better visualize the necessary areas in science (Piyatissa, Jahar & Tarofder, 2018).

Gilbert (2010) classified the above mentioned external representations into five modes, namely, (a) the concrete (the material) mode (physical models/practical setups); (b) the verbal mode (written and spoken expressions); (c) the symbolic mode (standard/scientific notations and mathematical equations); (d) the visual mode (diagrams, graphs, graphics, video clips, animations, video clips, etc., and (e) the gestural mode (teachers' gestures). Teachers usually use these external representation modes in classroom teaching. Though the external representations are used in the usual classroom teaching, paying more significant attention to how they can be used to elaborate the visualization aspects in science lessons is worthwhile (Piyatissa,

Jahar & Tarofder, 2018; Wong & Chu, 2017). In this context, identification of visualization related areas in science lessons, selecting suitable external representation mode(s) to elaborate the visualization aspect(s) and assessing the students' attainment of visualizing the respective areas are essential.

In this study, an attempt has been made to identify visualization related areas in selected physics lessons taught in secondary school level in Sri Lanka (students of aged 15–16 years in Grade 10 class), and design a lesson specific assessment tool to assess the students' attainment in visualizing those areas. This assessment tool is called Visualization Related Physics Test (VRPT). Further, selected computer-aided learning (CAL) materials such as video clips, animations, simulations, etc. were used in this study to elaborate visualization aspects in the lessons by reinforcing the usual classroom teaching. Physics is one component in the two years Integrated Science Curriculum taught in Grade 10 and Grade 11 classes in the Sri Lankan school system. The other components are biology and chemistry. The students sit for the General Certificate of Education (Ordinary Level) [GCE (O/L)] examination which is a national level evaluation, at the end of grade 11. Low physics performance of the students at GCE (O/L) examination as revealed by the evaluation reports issued by the Department of Examinations, Sri Lanka (for the years 2009–2015) is one reason for selecting physics for the study (over 50% of the students have scored marks in the range of 0–25% in the essay type questions). The lessons concerned in the study are kinematics, Newton's laws of motion, turning the effect of forces, force equilibrium, resultant of forces, frictional force, and mechanical energy. The reason for selecting these lessons in the grade 10 syllabus (National Institute of Education, 2015) is that they are the introductory lessons in many other physics courses and at the same time they are rich in visualization content too. However, similar studies need to be done for other physics lessons and the lessons in the biology and chemistry components as well.

In general, various visualization aspects can be noted in physics such as (a) building a mental image of an object physically seen, drawn, or heard about; (b) comparing physical quantities related with moving objects; (c) spatial imaginations of objects; (d) visualizing spatial relationships and orientations of objects; (e) visualizing algebraic equations; (f) mentally inverting, cross-sectioning or rotating an object;

(g) imaginations resulted from some hypothetical scientific models, etc. Therefore, selecting the most appropriate visualization aspect for elaborating specific subject content at the classroom teaching and assessing the students' familiarity with those aspects are, to a large extent, content-specific in nature. Because of this broad variety in visualization even in a particular lesson topic, finding a common framework for assessing students' attainment in visualization in physics is difficult. Therefore, a lesson and content-specific assessment tool were utilized here.

2. MATERIALS AND METHODS

2.1 Experimental Design

This study is a field experiment where an experiment is done in the existing school system. The experimental design used in this study is a 'pre-test and post-test experimental and control group design' (Sekaran & Bougie, 2013). Table 1 shows the experimental design.

Table 1: Experimental Design of the Study

Group	*Pre-test*	*Treatment*	*Post-test*
Experimental	(a) The Visualization Related Physics Test (V_1) (b) Physics questions at the end of Term1 school science term test paper (P_1)	Use of CAL materials to highlight the visual-perception, embedded with usual classroom teaching	(a) The Visualization Related Physics Test (V_2) (b) Physics questions in end of Term 2 school science term test paper (P_2)
Control	(a) The Visualization Related Physics Test V_3) (b) Physics questions at the end of Term1 school science term test paper (P_3)		(a) The Visualization Related Physics Test (V_4) (b) Physics questions in end of Term 2 school science term test paper (P_4)

Treatment effect on attainment in visualization = $[(V_2–V_1) – (V_4–V_3)]$
Treatment effect on general physics = $[(P_2–P_1) – (P_4–P_3)]$

As shown in Table 1, the experiment group is taught physics lessons by highlighting the visualization aspect using CAL materials, reinforcing the usual classroom teaching while the control group is also taught the same physics lessons of the same duration as for the experimental group intentionally not highlighting the visualization aspect with/ without using CAL materials. For both the groups, the Visualization Related Physics Test (VRPT) was administered as a pre-test and a post-test. Students' marks for the physics questions of the term-end school science papers (term 1 end: end of March 2017 and term 2 end: end of July 2017, that is before the experimental teaching and after the experimental teaching respectively) were extracted as a measure of their general physics performance. The random assignment of the students to the groups increased the internal validity of the experimental design. Selecting a sample of students from averagely performed schools increased the external validity of the experimental design.

During the time of the experiment, the participating students have completed the experimental lessons at their usual school time table. The experiment was conducted during the period, May–July 2017. The experimental lessons were conducted as after-school sessions. For both, the groups 15 hours of experimental teaching was done. One teacher conducted the experimental lesson for the experimental group while another teacher conducted the experimental lesson for the control group in each school. The medium of instruction was Sinhala. The CAL materials were used in the lessons with the experimental group as a whole class student participatory interactive demonstration sessions.

2.2 The Student Sample

A student sample of 184 was selected from three averagely performed government schools for the study. The composition of the sample is given in Table 2.

The whole experimental group and control group comprised of the students from all three schools. In this study, no school-wise differences or gender-wise differences were of interest. A sample comprising of a quite equal number of girls and boys has been achieved by selecting a girls' school, a boys' school and, a girls' and boys' mixed school.

Table 2: Composition of the Sample

School	*Participated Number of Students*	
	Experimental Group	*Control Group*
WP/Pili/Buddhist Girls' College Mount Lavinia (Girls School)	29	29
WP/Pili/Colombo South Science College, Mount Lavinia (Boys' school)	31	30
WP/Pili/Dehiwala S.de S. Jayasinghe Central College, Dehiwala (Mixed school)	32	33
Total	92	92

2.3 Notable Visualization Related Areas

Some of the notable visualization related areas of the above-mentioned physics lessons were identified after going through the grade 10 science syllabus and the teacher's guide (National Institute of Education, 2015), and the school textbook (Educational Publication Department, 2015). Table 3 gives a brief description of these areas.

Table 3: Outline of the Visualization Related Areas

Lesson	*Notable Visualization Related Areas*
Kinematics	(a) Differentiating the distance and displacement, and its impact on average speed and average velocity by visualizing the rectilinear motion of an object (treating plus and minus signs of the displacement and velocity) (b) Interpreting the acceleration concerning the distance or displacement (c) Interpretation of displacement-time and velocity-time graphs
Newton's laws of motion	(a) Visualizing the effect of an object kept on zero gravity, zero air pressure, and zero surface friction (background of Newton's first laws) (b) Visualizing Newton's first law of motion in terms of speed and the external force applied on the object (c) Visually predicting the effect of a constant force applied to an object continuously for a specific time (background of Newton's second law) in term of its speed

Table contd...

Table 3 contd...

Lesson	*Notable Visualization Related Areas*
	(d) Visualizing the effect of varying the variables in $F = ma$ in terms of the change in speed, of the object in question, keeping one variable constant at the time (Newton's second law of motion) (e) The imagination of action and reaction pair of forces under Newton's third law of motion
Turning effect of forces	(a) Visualizing the turning effect of forces regarding moments of the force (the magnitude of the force multiplied by the shortest distance to the line of action of the force from the point around where the moment is calculated) (b) Visualizing the principle of moments (clockwise moments are equal to anti-clockwise moments if the system is in force equilibrium)
Force equilibrium	(a) Visualizing the centre of gravity of the lamina regarding moments (b) Visualizing stability of an object (under the force equilibrium)
Resultant of forces	(a) Visualizing the resultant of parallel forces (b) Visualizing the direction of the resultant force of two angled forces
Frictional force	(a) Visualizing the factors affecting the frictional force (b) Visualizing the direction of the frictional force (c) Visualizing the static and dynamic friction by differentiating them
Mechanical energy	(a) Visualizing the definition of mechanical work (b) Visualizing kinetic and mechanical energy transformations
Free body diagrams for inclined planes*	(a) The clear general overall imagination of forces marked in a free body diagrams (b) Visualizing the direction of forces using free-body diagrams (c) Visualizing the effect of forces using free-body diagrams

*'Free body diagrams for inclined planes' is not included in the syllabus as a separate lesson. It is included here separately concerning its relation to other lessons.

Table 3 contains visualization aspects generally identified in this study. To elaborate them on classroom teaching, the science teachers can use various external representation modes. In this study, the CAL materials (coming under the visual mode) were used to improve the students' familiarity in visualizing these areas only for the experimental group.

2.4 Development and Validation of the Visualization Related Physics Test (VRPT)

2.4.1 *Distribution of Test Items in the VRPT*

The VPRT is a paper and pencil test consisting of a set of physics questions (test items) which can be considered as capable of measuring the students' attainment of visualizing the above mentioned notable visualization related areas of the physics lessons given in Table 3. The test consists of 28 multiple choice type questions which have to be answered within 30 minutes. The distribution of the test items in the VRPT is given in Table 4.

Table 4: Distribution of the Test Items in VPRT

Lesson	*Number of Questions (Number of test items)*
Kinematics	7
Newton's laws of motion	7
Turning effect of forces	4
Force equilibrium	2
Resultant of forces	2
Frictional force	2
Mechanical energy	2
Free body diagrams for inclined planes	2
Total number of items	**28**

The number of questions in each lesson was decided based on the duration of the lesson as mentioned in the grade 10 science syllabus (National Institute of Education, 2015).

2.4.2 *Sources of Test Items in the VRPT*

Nine test items of the test were developed by the researchers specifically for this study. The rest have been adopted/adapted from the existing sources. They are, science past papers of school term tests, science past papers in GCE-OL (Sri Lanka) examination, Physics past papers in Cambridge IGCSE Physics, O Level Physics MCQ Hot Spots 1000 Frequently Examined Questions - Red Spot Publications (Lim, 2015), IGCSE Physics Course Book-General Physics (Sang, 2011), and Mechanics Diagnostic Test (Korsunsky, 2003).

2.4.3 *Sample Test Items in the VRPT*

A special emphasis was made for elaborating the visualization aspects motioned in Table 3 using selected CAL materials, only for the students in the experimental group, special care was taken not to connect these CAL materials directly to these questions. This makes sure the control group students not being familiar with the new learning materials used with the experimental group do not experience any difficulty in understanding the questions. Three sample questions of the VRPT are given below.

Question 1

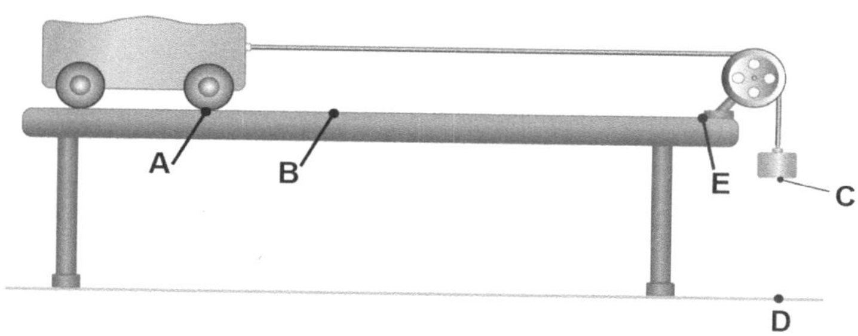

Figure 1: Experimental Setup for Sample Question 1

Figure 1, shows an experimental set up containing a trolley kept on horizontal smooth table. The trolley is attached a light string which goes through a smooth pulley fixed to the edge of the table and a weight is hung from the other end of the string. When the weight is released from the rest at C, it hits the ground at D.

AB = CD

Which of the following statements as shown in Table 5 is true regarding AB and BE motions of the trolley?

This question assesses the students' attainment in 'Visually predicting the effect of a constant force applied to an object continuously for a specific time (background of Newton's second law) in terms of its speed' to predict the motion of the trolley in AB. To predict the motion of the trolley in BE, the ability in 'Visualizing Newton's first law of motion in terms of speed and the external force applied to the object' is needed.

Table 5: Multiple Choices for Question 1

	AB	*BE*
(a)	Acceleration	Uniform velocity
(b)	Acceleration	Acceleration of magnitude which is less than that in AB
(c)	Acceleration	Deceleration
(d)	Uniform velocity	Uniform velocity of magnitude which is less than that in AB

Question 2

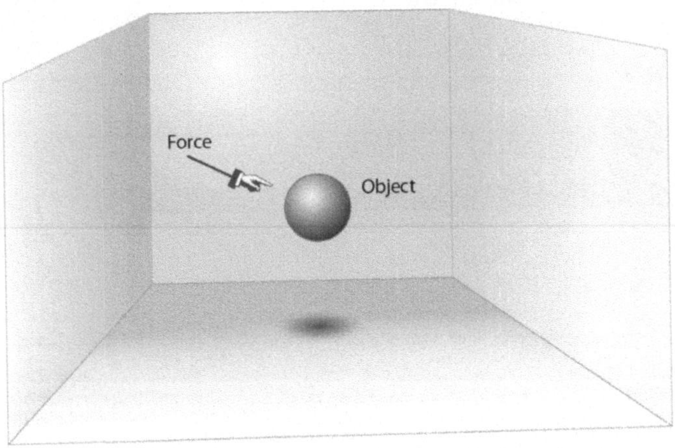

Figure 2: Experimental Setup for Sample Question 1

Figure 2 shows a cross sectional view of room under idealized conditions: it is in under zero gravity, zero air pressure (no air resistance) and, all the surfaces are frictionless. Assume that an object is placed inside the room in the space (without touching any surface of the room) as shown in the diagram. A push is given to the object into the shown direction and then the push is released. Three statements are given regarding the situation.

I. Before applying the force it will stay at the same initial position,
II. It will travel with the velocity it attained when the push is released until it hits on a surface of the room.

III. After hitting on a surface of the room also its speed will remain unchanged

Which of the above statement(s) is/are correct?

(a) Only I
(b) Only II
(c) Only I and II
(d) All the three.

The major visualization related area this question assesses is 'Visualizing the effect of an object kept on zero gravity, zero air pressure, and zero surface friction (Newton's first law)'.

Question 3

A long truck has gone out of the road and fallen into a ditch. A and B breakdown vehicles are used to take out the truck from the ditch as shown in Figure 3.

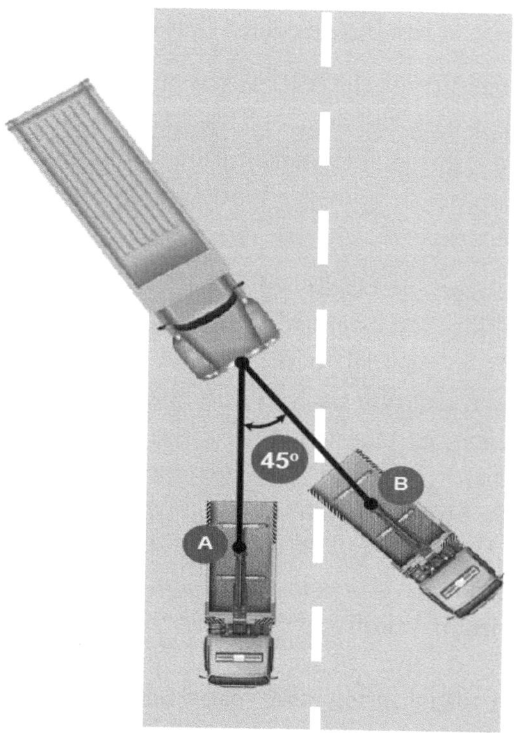

Figure 3: Experimental Setup for Sample Question 3

When the truck is pulled by A and B vehicles using the strings as shown in the diagram, the truck may move forward:

(a) along OA
(b) along OB
(c) along a direction between OA and OB
(d) along a direction out of BOA angle.

This question assesses the attainment in 'Visualizing the direction of the resultant force of two angled forces'.

2.4.4 *Reliability and Validity of the VRPT*

The reliability of the VRPT was tested in a pilot study conducted with 67 students of the three schools not participated in the main experiment and a Cronbach's alpha value of 0.716 was obtained. In the pre-test administration of the VRPT in the main experiment, a Cronbach's alpha value of 0.711 was obtained. For a classroom assessment, if this value is around 0.700, it can be concluded that the assessment is reliable or having the internal consistency (Santrock, 2009). Therefore, it can be considered that the VRPT is a reliable tool.

The content validity of the VRPT was tested. That is, whether the questions match with the depth of the grade 10 science syllabus and they contain a considerable visualization component was assured by getting the comments from the participating science teachers and a group of subject experts.

To test the concurrent validity of the VRPT, the aggregate scores of three spatial ability tests were administered for participating students. This was done in parallel to the pre-test administration of VRPT. The three spatial ability tests used were the Paper Folding Test (PFT) (Ekstrom, French, & Harman, 1976), the Perspective Taking/Spatial Orientation Test (PTSOT) (Hegarty & Waller, 2004), and the Revised Perdue Spatial Visualization Tests: Visualization of Rotations (Revised PSVT: R) (Guay, 1976; Yoon, 2011). The major reason for selecting the students' scores in spatial ability as a measure suitable to test the concurrent validity of VRPT is the strong literature support that the ability to solve science (physics) problems involving an overt component of visualization and spatial ability are positively related (e.g. Gilbert, 2005; Kozhevnikov *et al.*, 2007; Lyna & Fulmer, 2014).

The correlation coefficient (Pearson's) of 0.526 (p-value = 0.000) between the spatial ability score and the VRPT score showed that the VRPT can be considered as having the concurrent validity.

To test the predictive validity of the VRPT, the correlation between the pre-test scores of the VRPT and the school term test physics marks were calculated. The correlation coefficient value (Pearson's) of 0.802 (p-value = 0.000) showed that the VRPT has predictive validity.

The construct validity is the extent to which there is evidence that the VRPT measures an unobservable trait, the students' level of visual-perception in physics. The judgments about construct validity might also rely on a description of the development of the tool (test) (Santrock, 2009). In developing the VRPT, the test items which are capable of measuring the students' familiarity with the visualization aspect in physics had been carefully selected. Therefore, it can be said that the VRPT has construct validity to a certain extent. However, in this study, no factor analytic type of analysis was done for testing the construct validity of the VRPT as the test had not been developed based on a factor structure due to the broad variety in visualization even within a particular lesson.

It can be concluded that the VRPT which is a lesson specific assessment tool developed in this study is a reliable and valid assessment tool suitable to measure the students' attainment in visualizing the necessary areas in physics.

2.5 Highlighting the Visualization Aspect through CAL Materials

How the selected CAL materials were used to elaborate on the visualization related areas are discussed below with three examples. Most of the materials used were downloaded from the World Wide Web. They are available to the end-user free of charge. These materials were selected after closely evaluating them on whether they are relevant to the grade 10 science syllabus (National Institute of Education, 2015) (considering the learning outcomes in the syllabus and limitations guided by the teacher's guide). These CAL materials can generally be considered as supporting materials developed by their producers/presenters for secondary school introductory physics courses.

In this study, they are used as visualization tools to help the students to build necessary visualizations.

2.5.1 *Kinematic Graphs*

According to the grade 10 science syllabus (National Institute of Education, 2015), kinematics graphs include displacement-time graphs and velocity-time graphs. Kinematics graphs involve visualization of moving objects and relating them with the graphs where the ability of visualization can be considered as very important. To introduce the displacement-time graphs, a CAL program developed by Tiros (https://www.youtube.com/watch?v=UefWw5k4G0U) was used. In this program, the displacement-time graph is drawn when the ball moves on the ruler in real-time (Figure 4).

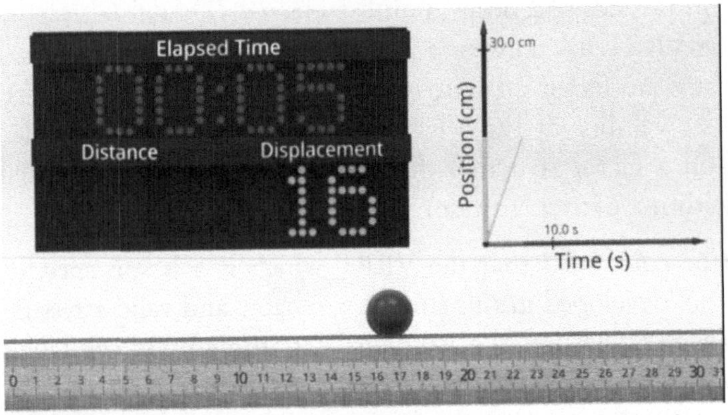

Figure 4: Real-Time Drawing of a Displacement-Time Graph
(*Source:* https://www.youtube.com/watch?v=UefWw5k4G0U)

The primary advantage of using this program is that it shows the movement of the physical object, changes in numerical values of displacement and time, and the displacement (position)-time graph in one screen. Several screens of the program explaining how the changes in displacement of a ball moving on a meter ruler relate to a displacement-time graph were used with teacher explanations. Then converting a displacement-time graph into the corresponding velocity-time graph was explained using the program. This program can be considered as helping the students to visualize the relationship between a displacement-time graph and the corresponding velocity-time graph by showing them the real-time visual effects of the two cases.

2.5.2 *Newton's Laws of Motion*

Newton's laws of motion can be considered as a section of physics in the grade 10 science syllabus (National Institute of Education, 2015) where the images of still and moving objects in relation with the direction of force inserted on the object and the direction of the motion of the object have to be built by the students where the ability to visualize can be considered as necessary in coping up with the subject matter with much ease.

An interactive simulation (*PhET*) called 'Force and Motion Basics' (https://phet.colorado.edu/sims/html/forces-and-motion-basics/latest/forces-and-motion-basics_en.html) was used to elaborate Newton's first and second laws of motion. Under 'Force and Motion', 'Motion' submodule was used here. In this interactive simulation, the student can decide the force applied by the man and keep the force at a constant value by holding the mouse, and then how the speed increases can be seen in the speedometer on top of the screen (Figure 5).

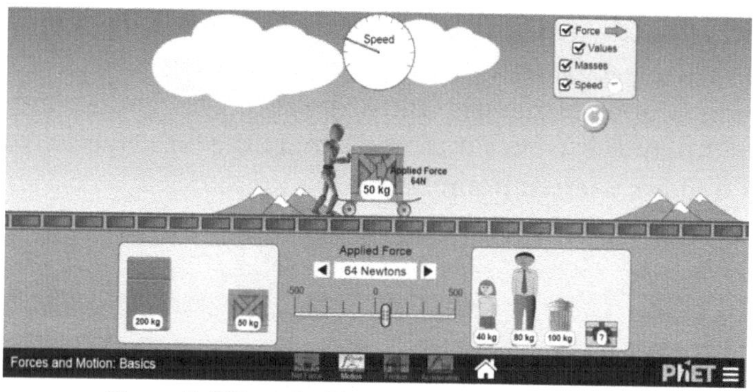

Figure 5: When the Force is Applied the Speed is Displayed in Real Time on the Speedometer

(*Source:* https://phet.colorado.edu/sims/html/forces-and-motion-basics/latest/forces-and-motion-basics_en.html)

Then the student can change the force applied to the cart (object) and can change the mass of the object. The speedometer on top of the screen shows the change in speed. The student can visualize the effect of a force applied to the object and mass of the object. This interactive lesson was done as a whole class interactive demonstration. Further, the above 'Motion' simulation can be used to help the students to build imaginations in Newton's first law of motion also. There is an option in this program for converting the moving track into a frictionless surface. Then, once the force applied to the object by the man is released by releasing the mouse, the object must move forward with the constant velocity. Now there is no net force, and the object must move forward on the track with the same velocity is attained at the moment the force is released.

2.5.3 *Turning Effect of Forces*

Under the grade 10 science syllabus (National Institute of Education, 2015), turning effects of forces involve the introduction of the concept of moment of forces and application of the principle of moments for simple systems. Understanding the principle of moments needs the visualization of the turning effect caused by the forces and determining how the forces cause the deviations from the equilibrium state of the whole system.

As an effort of helping the students to visualize the application of the principle of moments, an interactive PHET simulation called 'Balancing Act' downloaded from https://phet.colorado.edu/sims/html/balancing-act/latest/balancing-act_en.html was used. The simulation consists of three modules 'Intro', Balance Lab', and 'Game' where it takes the student from simple tasks to a little bit complicated tasks. Figure 6 shows a screen from 'Intro'.

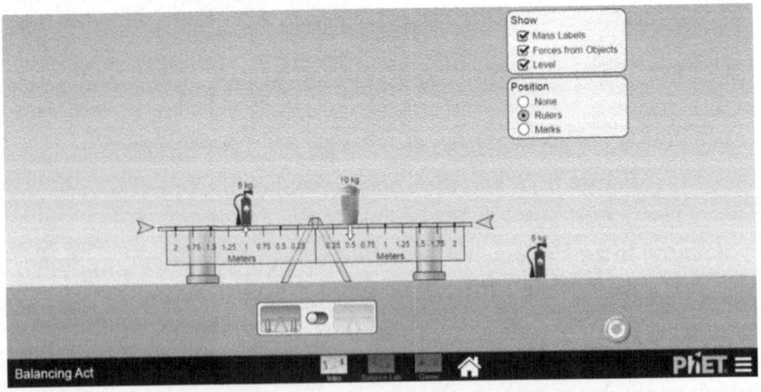

Figure 6: Balancing the Weights on the Seesaw
(*Source:* https://phet.colorado.edu/sim s/html/balancing-act/latest/balancing-act_en.html)

In Figure 6, the student can move the weight here and there on the seesaw and by removing the two side supporters whether the seesaw is balanced or not can be seen. By repeating the same thing using different weights the student can derive the principle of moments, that is, total clockwise moments are equal to the total anti-clockwise moment if the system is at the force equilibrium. In 'Balance Lab', the student can experience the same thing by using several masses. 'Game' module presents some interesting problem-solving tasks.

2.6 Methods of Statistical Analysis

For testing the reliability of the VRPT, Cronbach's alpha value was calculated. The Coefficient of Correlation (Person's) was used in testing the concurrent validity and the predictive validity of the VRPT. Descriptive statistics such as mean, standard deviation and skewness have been calculated where necessary.

It is expected that the students in the experimental group having exposed to the experimental teaching emphasizing the visualization aspect should indicate a notable gain in the post-test performance in the VRPT compared to their pre-test performance. It is also expected that similar gain to the gain as mentioned above in the control group, if any, should be lower compared to that in the experimental group as the control group has only been exposed to the standard teaching without emphasizing the visualization aspect. To statistically test these effects, the group comparison tests were used.

3. RESULTS AND DISCUSSION

The results of the group comparison tests are discussed here. Testing the reliability and validity are not discussed here as they have been discussed under '2.4 The Development and Validation of the VRPT'. Table 6 and Figure 7 show the mean of total scores obtained by the students of the two groups for the VRPT at the pre-test and the post-test.

Table 6: Students' Performance in VRPT (mean total scores as percentages)

Group	*Pre-test*		*Post-test*	
	Mean	*Std. dev.*	*Mean*	*Std. dev.*
Experimental	32.11%	13.32%	50.13%	13.80%
Control	32.17%	14.95%	33.78%	14.78%

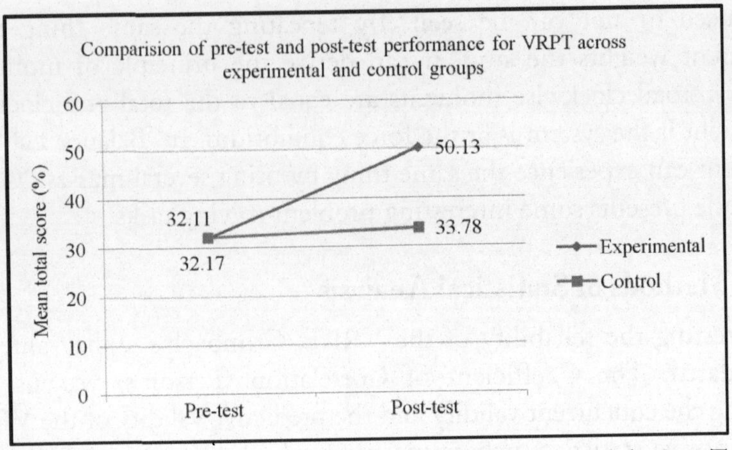

Figure 7: Students' Performance in VRPT in the Pre-Test and the Post-Test

According to Table 6, the sample shows a quite high standard deviation. Table 6 shows that the performance after the treatment has gone up in the experimental group by 18.02% and the improvement in the control group is only 1.61%. Both the improvements are statistically significant as the p-value of the Wilcoxon Signed Rank Test is less than 0.05 (Table 7). The initial performance (pre-test) is almost similar between the two groups as the p-value is greater than 0.05 for the Welsh test (Table 8). Further, there is a 16.41% increase in the post-test performance in the experimental group compared to the control group which is statistically significant as the p-value from the Welsh test is less than 0.05 (Table 8).

Analysis of the students' performance for the individual test items of the VRPT supports the above analysis. Figure 8 compares the students' performance for VRPT between the pre-test and the post-test of the experimental group and Figure 6 compare tat for the control group.

Table 7: Testing the Significance of the Difference between the Pre-test and the Post-test Total Mean Scores within a Group (Related group comparison results)

Group	*Mean Pre-test Total Score*	*Mean Post-test Total Score*	*Difference*	*p-Value from Kolmogorov-Smirnov Test of Normality*	*p-Value from Wilcoxon Signed-Rank Test*
Experimental	32.11%	50.13%	18.02%	0.000	0.000
Control	32.17%	33.78%	1.61%	0.000	0.000

Table 8: Testing the Significance of the Difference between Total Mean Scores of the Experimental Group and the Control Group for the Pre-test and the Post test (Independent group comparison results)

Test	*Mean Total Score (%)*		*Mean Difference*	*p-Value form Levene's Test for Equality of Variance*	*Skewness*		*p-Value from Welsh Test*
	Experi-mental Group	*Control Group*			*Experi-mental Group*	*Control Group*	
Pre-test	31.11	32.17	1.06	0.011	0.437	0.287	0.964*
Post-test	50.13	33.78	16.41	0.031	0.466	0.247	0.000

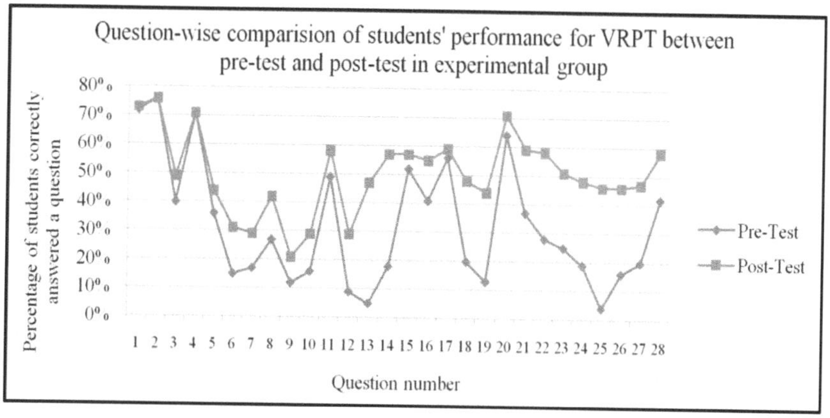

Figure 8: Question-wise Comparison of VRPT Performance in Experimental Group

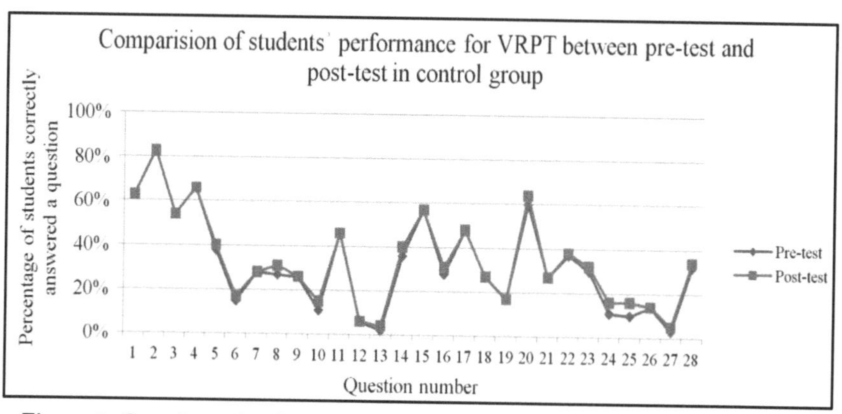

Figure 9: Question-wise Comparison of VRPT Performance in Control Group

As shown in Table 9, there is a 14.41% improvement in the general physics performance in the experimental group. Similarly, that improvement for the control group is 4.16%. Further, there is a 10.25% improvement in the general physics performance in the experimental group compared to the control group. Statistical information regarding the significance is given in Table 9 and Table 10.

Table 9: Comparison of the Term 1 End and Term 2 End Physics Performance before and after the Treatment in the Same Group

Group	*Mean Term 1 Physics Score (%)*	*Mean Term 2 Physics Score (%)*	*Difference Between*	*p-Value from Kolmogorov-Smirnov Test of Normality for the Mean Difference*	*p-Value from Wilcoxon Signed-Rank Test*
Experimental	35.09	49.50	14.41	0.000	0.001
Control	35.23	39.39	4.16	0.000	0.002

Table 10: Comparison of the Term 1 end and Term 2 End Physics Performance before and after the Treatment Across the Two Groups

Term End Test	*Mean Physics Score (%)*		*Mean Difference*	*p-Value form Levene's Test for Equality of Variance*	*Skewness*		*p-Value from Welsh Test*	*p-Value from Mann-Whitney Test*
	Expt. Group	*Control Group*			*Expt. Group*	*Control Group*		
Term 1 end	35.09	35.23	0.14	0.026	.293	–.286	–	0.805*
Term 2 end	49.5	39.39	–10.11	0.003	.252	–.227	–	0.000

As shown in Table 9, the p-values for the Welsh test is below 0.05 for both the groups. Therefore, there is a significant difference between the Term 1 end physics performance and Term 2 end physics performance for both the groups. As shown in Table 10, for the Term 1 end physics performance there is no significant difference between the experimental and the control groups as the p-values from the Welsh test is more than 0.05. For the Term 2 end physics performance, there is a significant difference between the experimental and control groups, as the p-value from the Welsh test is less than 0.05. Therefore, it can be concluded that the teaching approach tested with the experimental group resulted in improving general physics performance.

4. CONCLUSIONS

This experimental study is an attempt of emphasizing the visualization aspect of the selected secondary school physics lessons, kinematics, Newton's laws of motion, turning the effect of force, force equilibrium, resultant of forces, frictional force and mechanical energy, basically utilizing the visual-mode of external representations (Gilbert J., 2010). Computer-aided learning materials such as video clips, animations, and simulations were used with verbal explanations in this study to supplement the usual classroom teaching.

The significant gain in VRPT scores of experimental group students (16.41%) showed over the control group students show that the new approach, highlighting the visualization aspect in physics, is capable of improving the students' attainment in visualizing the principles in physics which are not very tangible to senses. Further, the gain of 10.25% in general physics performance of the experimental group over the control group showed that the teaching approach tested with the experimental group resulted in improving the students' general physics performance as well.

It is recommended that for other units of secondary school physics such as hydrostatics, electricity, waves, optics, electromagnetism, electronics, etc. and also for the sections of the other component subjects in science such as biology and chemistry, it is worthwhile to apply the new approach, which will be helpful for popularizing the science education.

However, it should be mentioned that using CAL materials is only one method of elaborating this aspect and in usual classroom teaching the teacher may use other suitable methods (Bybee, 2014; Osborne, 2014) as well which could strengthen students' visualizations, internal representations or imaginations. Some of them are the use of practical setups, verbal explanations by way of similes/story-telling, annotated diagrams, 3D diagrams, etc. Whatever the materials the teacher uses the most important thing is directing the students to visualize the necessary areas in science. In this respect, the teacher has to identify the important visualization related areas in the lessons before conducting the lesson.

REFERENCES

[1] Bybee, R. (2014). NGSS and the next generation of science teachers. *Journal of Science Teacher Education, 25*(2), 211–221.

[2] Department of Examinations Sri Lanka (2009). *Evaluation Report, G.C.E. (O/L) Examination: 34-Science.* Battaramulla, Sri Lanka.

[3] Department of Examinations Sri Lanka (2010). *Evaluation Report, G.C.E. (O/L) Examination: 34-Science.* Battaramulla, Sri Lanka: Department of Examinations.

[4] Department of Examinations Sri Lanka (2011). *Evaluation Report, G.C.E. (O/L) Examination. Department of Examinations: 34-Science.* Battaramulla, Sri Lanka.

[5] Department of Examinations Sri Lanka (2013). *Evaluation Report, G.C.E. (O/L) Examination. Department of Examinations: 34-Science.* Battaramulla, Sri Lanka.

[6] Department of Examinations Sri Lanka (2014). *Evaluation Report, G.C.E. (O/L) Examination: 34-Science. Department of Examinations.* Battaramulla, Sri Lanka.

[7] Department of Examinations Sri Lanka (2015). *Evaluation Report, G.C.E. (O/L) Examination: 34-Science. Department of Examinations.* Battaramulla, Sri Lanka.

[8] Department of Examinations Sri Lanka. (2012). *Evaluation Report, G.C.E. (O/L) Examination. Department of Examinations: 34-Science.* Battaramulla, Sri Lanka.

[9] Educational Publication Department (2015). *Grade 10 science textbook.* Battaramulla, Sri Lanka.

[10] Ekstrom, R., French, J. and Harman, H. (1976). *Manual for kit of factor-referenced cognitive tests* (Vol. 102). NJ: Educational Testing Services.

[11] Gilbert, J. (2010). The role of visual representations in the learning and teaching of science: An introduction. *Asia-Pacific Forum on Science Learning and Teaching,* 11(1).

[12] Gilbert, J.K. (2005). Visualization: A Metacognitive Skill in Science and Science Education. In J.K. Gilbert, *Visualization in Science Education* (pp. 9–27). Dordrecht: Springer.

[13] Guay, R.B. (1976). *Purdue Spatial Visualization Test.* West Lafayette: Purdue Research Foundation.

[14] Hegarty, M. and Waller (2004). A dissociation between mental rotation and perspective-taking spatial abilities. *Intelligence, 32*(2), 175–191.

[15] Korsunsky, B. (2003). *Mechanics Diagnostic Test.* Harvard Graduate School Education.

[16] Kozhevnikov, M., Motes, M. and Hegarty, M. (2007). Spatial Visualization in physics problem solving. *Cognitive Science, 31,* 549–579.

[17] Lim, C. (2015). O Level Physics MCQ Hot spots 1000 Frequently Examined Questions. *Red Spot Publications.*

[18] Lyna and Fulmer, G. (2014). Secondary students' visual-spatial ability predicts performance on the Visual-Spatial Electricity and Electromagnetism Test (VSEEMT). *Science Education Review Letters*, 8–21.

[19] National Institute of Education (2015). *Grade 10 science teacher's guide.* Sri Lanka: Educational Publications Department of Sri Lanka.

[20] Osborne, J. (2014). Teaching scientific practices: meeting the challenge of change. *Journal of Science Teacher Education, 25*(2), 177–196. doi: 10.1007/s10972-014-9384-1.

[21] Piyatissa, M., Jahar, M. and Tarofder, A. (2018). Impact of visualization on teaching and learning physics at secondary school level. *European Journal of Education Studies,* 4(6), 114–144.

[22] Reiner, M. (2008). Seeing through touch: The role of haptic information in visualization. In M.R. Gilbert and M. Nakhleh, *Visualization: Theory and Practice in Science Education* (3 ed., pp. 73–84). Netherlands: Springer.

[23] Sang, D. (2011). *Cambridge IGCSE Physics Course Book.* UK.: Cambridge University Press.

[24] Santrock, J. (2009). *Educational Psychology* (4 ed.). NY: McGraw-Hill.

[25] Sekaran, U. and Bougie, R. (2013). *Research Methods for Business—A Skill Building Approach* (6 ed.). U.K: John Wiley & Sons Inc.

[26] Shu-Nu, C.-R. and Yao, B.J. (2014). *Visualization in research and science teachers' professional development. Asia-Pacific Forum on Science Learning and Teaching* (Vol. 15). Hong Kong Institute of Education.

[27] Wong, C. and Chu, H.-E. (2017). The conceptual elements of multiple representations: A study of textbooks' representations of electric current. In D.F. Treagust, D.R. and H.E. Fischer, *Multiple Representations in Physics Education. Models and Modeling in Science* (pp. 183–206). New York: Springer.

[28] Yoon, S. (2011). *Psychometric properties of the revised Purdue spatial visualization tests: visualization of rotations (The revised PSVT: R).* West Lafayette, Indiana: Doctoral dissertation, Purdue University.

Promoting of Science and Technology through Story Narration and Displaying Science History with Animation

Hassan Bashiri[1] and Amir-Hushang Heidari[2]

[1]Department of Computer Science, Hamedan University of Technology, IRAN

[2]Department of Futures Studies, National Institute for Science Policy, IRAN

E-mail: [1]bashiri@hut.ac.ir

ABSTRACT: Science history is full of wonders. Newton's unceasing passion for describing the world's mechanism as an exact clock and the discovery of gravity is an incredible story as well as the invention of differential equations along with the real challenges of Newton's life. Einstein's great endeavors to understand physics and the use of mathematics, ethical challenges in his personal life, and his political attempts to influence politicians narrates the image of a scientist who is not just intelligent and a genius, but like everyone has a life full of challenges and pressures. There are many stories in the history of science and technology which can motivate, inspire and sometimes solve the challenges of today's science. In most cases, the history of science has been presented as a technical text to enthusiasts. A structure that has less impact on the insight of students and learners.

In this study, a framework was defined by asking which method has more impact on students and learners, which presents better the evolution of science history and which is more inspiring. The purpose of this study is to investigate the effectiveness of story narration and animation in the teaching of science history in physics, astronomy, mathematics, and medicine. For this purpose, Wikipedia articles were first provided to about 50 students in the mentioned topics, and they answered 20 questions after studying. A total of 50 other students were selected with the same average in their courses as the first group, and the same material was presented as a science documentary and animation. Again, these students were asked about 20 questions. The mean score of the second group was 25% higher than the first group. After three months, the same procedure was repeated for both groups to determine the sustainability of training. The first group had 20% lower scores than the first test, while the second group experienced only a 5% reduction.

This study suggests that, in addition to learning more using story narration and animation for students, the sustainability of teaching in the second method is also far better. It should be noted that we used the "Cosmos: A Spacetime Odyssey" documentary and for technology stories we used the scientific documentary "From Dream to the Invention".

Keywords: Story Narration, Scientific Animation, Promotion of Science and Technology.

1. INTRODUCTION

Science education has faced serious challenges for a wide range of students. Displaying science and promoting it in the structure of the text even with some progress in traditional media, is still boring for students. Various studies have been conducted by researchers on the effectiveness of different teaching methods (Ainsworth & Loizou, 2003; Ainsworth & VanLabeke, 2004; Ayres & Paas, 2008). Research has also been done to investigate high school students' feedback on various educational methods as a doctoral dissertation by Smith at the University of Iowa (Reid-Smith, 2013). Besides, the availability of information and educational resources has also increased dramatically over the past decade.

2. PROBLEM STATEMENT

There are a variety of educational resources available on different websites, including text, images, tutorials, online lessons, and videos. With the creation of platforms for uploading educational resources, the participation of educators in producing content has also increased incredibly. These include online courses offering businesses such as Udacity, Coursera, Lynda, Iversity, Udemy, Codeacademy, Khanacademy, and educational content delivery platforms such as faradars.org and maktabkhooneh.org for Persian audiences. Self-learning has also become one of the key skills. Learning how to learn means finding the most effective resources for learning and mastering the issue in the shortest time (Pigozzi, 2006; da Costa Santos &

Quiroga, 2019). In many cases, the thirst for learning technologies and its tools has increased so much among educators and practitioners that they have neglected the process of creating the right insights for the proper application of these tools and technologies. Continuing this process will have a negative impact on innovation and creativity. What is needed in the large volume of information and educational resources is a deep insight into and understanding of scientific methods and processes.

The history of science is full of surprises. Exciting wonders that give enthusiasts the insight to keep going. Newton's boundless passion for describing the world like a clock and discovering gravity and devising differential equations alongside the real challenges of Newton's life is an interesting story. A narrative that ties together curiosity and constant effort. Invention stories in the past and present and narrative of successful startups are inspiring. An in-depth insight is created by diving into the past and an analytical understanding of the scientific mechanism and examining of the solutions used by ancestors to solve problems (Allchin, 2013).

3. SUGGESTED SOLUTION

Given our personal experience of the effectiveness and persistence of the teachings based on fictional and animated narratives, we organized a study to investigate the effectiveness of deep insight using animation and narrative science history. In search of visual and animated sources, we chose the documentary series "Cosmos: A Spacetime Odyssey" and for technology stories, we used the scientific documentary "From Dream to the Invention".

Figure 1 shows the story of Edmond Halley's curiosity about the comet when the plague swept Europe and London was on fire, and the comet's presence was a scary and miserable message to all people for years to come. It is a fascinating narrative from the questioner's point of view of a teenager, which was not terrifying but full of charm.

Figure 2 shows the narrative of Newton's discovery of a white light structure that combines several colors. The moment the spectrum of light was discovered by a prism that put Newton in front of white light. Newton gave up when he discovered the color spectrum. But William Herschel completed Newton's work with one question. Do some light colors have more heat? The essence of a science genius is in asking what the rest of us have accepted without question, and this is when Herschel came up with a rigorous experiment to answer his question in the infrared light (Figure 3).

Figure 1: Edmund Holly's Curiosity Narrative

Figure 2: The Moment of Newton's Discovery of the Spectrum of Light

Figure 3: The Moment of William Herschel's Discovery of the Infrared Spectrum

4. EXPERIMENT DESIGN

Based on the interesting and narrative sections in the two mentioned documentaries, questions were drawn on the theoretical foundations of the sciences that represent a profoundly instructive understanding. Texts that cover the issues raised were extracted from Wikipedia. Two groups of 50 students each interested in physics, astronomy, mathematics, and inventions were selected. The educational average of both groups was approximately equal. The first group studied the texts and answered 20 questions. The second group watched and recorded selected portions of the documentaries. The same questions were given to the second group. The mean scores of the second group were 25% higher than the first group.

The same procedure was repeated for another three months to determine the sustainability of training. The first group had a 25% reduction in scores compared to the first test, while the second group experienced only a 5% decrease. Average scores obtained by both groups in the two rounds are presented in Figure 4. The blue color represents the first stage of both groups and the orange color of the second stage test, which was held three months later with the same questions from both groups.

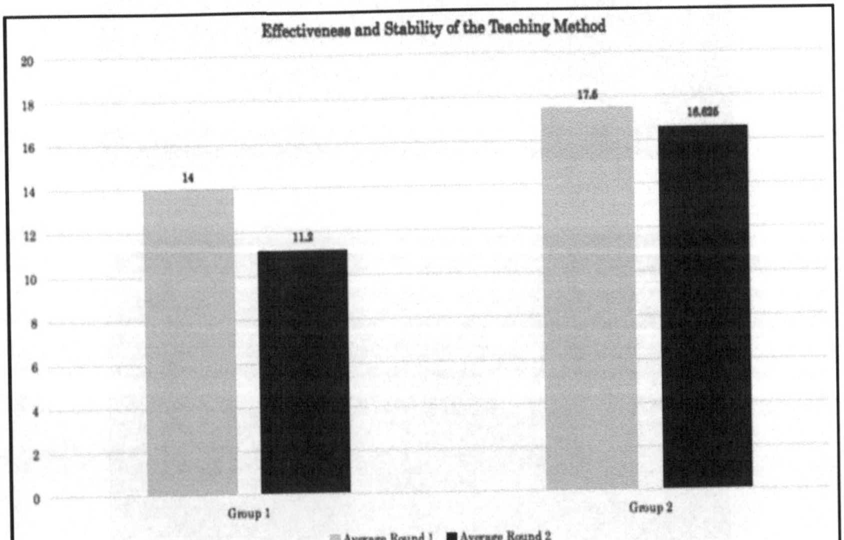

Figure 4: Efficacy and Sustainability of Text-Based and Narrative-Animated Teaching Methods

5. CONCLUSION

This research shows that creating scientific attraction using new technologies and applying narrative animation in addition to better learning for students has higher educational sustainability. Educational materials for the present research were selected from the documentary series "Cosmos: A Spacetime Odyssey" and for technology stories, we used scientific documentary "From Dream to the Invention".

REFERENCES

[1] Ainsworth, S. and Loizou, A.T. (2003). The effects of self-explaining when learning with text or diagrams. *Cognitive Science,* 27(6), 937–937.

[2] Ainsworth, S. and VanLabeke, N. (2004). Learning and Instruction. *Multiple forms of dynamic representation* , 14(3), 241–255.

[3] Allchin, D. (2013). *Teaching the nature of science: Perspectives and resources.* Saint Paul: MN: SHiPS Education Press.

[4] Ayres, P. and Paas, F. (2008). Learning from animations: A cognitive load approach. *Manuscript submitted for publication.*

[5] da Costa Santos, J. and Quiroga, M.L. (2019). UNESCO's Contributions to the Development of Education in Latin American Countries: Brazil and Bolivia. *2nd International Conference on Education Science and Social Development (ESSD 2019).* Atlantis Pres.

[6] Pigozzi, M.J. (2006). A UNESCO view of global citizenship education. *Educational Review, 58*(1), 1–4.

[7] Reid-Smith, J.A. (2013). *Historical short stories as nature of science instruction in secondary science classrooms: Science teachers' implementation and students' reactions.* PhD Thesis, Iowa State University.

Impact of Popularization of Science on Knowledge-Based Economy Indicators

Nasser Ali Azimi* and Elham Fakhremoosavi
Department of Economics of Science,
National Research Institute for Science Policy, Tehran, IRAN
E-mail: *naazimi5@yahoo.co.uk

ABSTRACT: Nowadays, the move to a knowledge-based economy where economic growth is driven by the production of science and innovation is of paramount importance to the countries' comparative advantage and their growth and development. In the knowledge-based economy, the application of knowledge and information is of the utmost importance and the production and distribution are based on knowledge. There are several methods used to measure the knowledge base of an economy, the most important being the World Bank's knowledge assessment methodology. Based on this framework, the successful transition to a knowledge-based economy requires long-term strategies for investing in the four pillars of the knowledge-based economy. These four pillars are education and human resources, effective innovation system, information infrastructure, economic incentive and institutional regime. This paper, with an analytical and pathological perspective, examines some of the variables of knowledge-based economy indicator, according to the role of popularization of science. It concludes that the quantitative criteria selected in these pillars, regardless of the quality of popularization of science in the countries, does not provide an accurate picture and comparison of countries.

Keywords: Knowledge-Based Economy, Popularization of Science, Developing and Developed Countries.

1. INTRODUCTION

Considering the role of knowledge in economics and economic growth is not new. In previous economic theories, science and technology have always been an important topic in economic growth theories. Knowledge-based economy is a new economic model that is becoming increasingly important, due to the development of ICT and its impact on the process of economic development and development of countries of various dimensions. Since in the knowledge-based economy science, technology, innovation, and entrepreneurship are the factors influencing economic growth, knowledge is regarded as the main type of capital and economic growth will be rooted in the accumulation of knowledge. For many economists, today the size of the capital and the size of the market no longer play a central role in the economic development of nations, but it is the knowledge and technology that plays this role. Accordingly, the right criteria should be used to measure the degree of knowledge that an economy has. Various organizations have proposed various methods to measure the knowledge base of an economy. The most famous and widely used of these is the World Bank's "Knowledge Assessment Methodology.

From Joseph Stiglitz's view, knowledge is a universal public good and will have the greatest impact on society once it is distributed without neglect (Stiglitz, 1998). One of the most important ways to disseminate science and knowledge in society is to promote science. It is done in any society different from other societies. The titles of "information society" and "knowledge-based society" mean that information and knowledge penetrate all social layers and do not take a step except by relying on relevant scientific information. Promoting science is also an attempt to realize ideal societies known as "information" or "knowledge-based" traits. The knowledge society is a society in which information is circulated among people in addition to databases, enabling individuals to be the creators of knowledge.

The question now is, to what extent can the World Bank's Knowledge Assessment Index be a strong estimator of the distribution of science in society, and can it also measure the quality of science distribution in societies? Given the necessity of understanding the knowledge-based economy and achieving its principles, we first examine the concept of the knowledge-based economy and its consequences. In the next section, we discuss the importance of the popularity of science and the role it can play in the transition to a knowledge-based economy. We then compare the experiences of some successful and unsuccessful countries in scientific popularity and finally make suggestions.

2. KNOWLEDGE-BASED ECONOMY

The term knowledge-based economy was first coined by the OECD in 1996 for those economies that are directly based on the production, distribution, and use of knowledge and information. According to the OECD, the knowledge economy is an economic foundation formed by the production, distribution, and application of knowledge. In this regard, information and investment in knowledge and knowledge-based industries are of particular interest.

The APEC Economic Committee considers knowledge-based economics an economy in which the production, distribution, and application of knowledge are the main drivers of economic growth, wealth production and employment in all industries. According to this definition, the knowledge-based economy for wealth growth and production does not rely solely on some high-tech industries, but in that economy, all industries use knowledge to meet their needs.

Table 1: KAM Basic Scorecard Variables

Economic incentive and institutional regime
Tariff and nontariff barriers
Regulatory quality
Rule of law
Education and human resources
Adult literacy rate (% age 15 and above)
Secondary enrollment
Tertiary enrollment
Innovation system
Researchers in R&D, per million population
Patent applications granted by the USPTO, per million population
Scientific and technical journal articles, per million population
Information infrastructure
Telephones per 1,000 persons (telephone mainlines + mobile phones)
Computers per 1,000 persons
Internet users per 10,000 persons

Source: Chen & Dahlman, 2005.

There are several methods used to measure the knowledge base of an economy, the most important of which is the World Bank's knowledge assessment methodology. This methodology introduced by Chen & Dahlman (2005) provides a framework to facilitate the transition of countries to a knowledge-based economy. Under this framework, the successful transition to a knowledge-based economy typically requires long-term investment strategies in the four axes of a knowledge-based economy, such as training and human resources, development of innovation capabilities, modernization of information and communication infrastructure, and having a favorable economic environment for market transactions. These axes are recognized by the World Bank as the foundations of the knowledge economy and constitute the framework of a knowledge-based economy. Table 1 shows knowledge-based economy variables related to each of the pillars.

3. POPULARIZATION OF SCIENCE

Popularization of science is the shortest way to achieve scientific results, while at the same time reinforcing our understanding of the facts. The goal of popularization of science is first to stimulate human attention to science and to present the most important scientific results in a very simple but realistic way. The dissemination and popularization of science if sustained and institutionalized can enhance scientific thinking and rationality in society. It paves the way for the advancement of science and its institutionalization and ultimately the progress and prosperity of society (Ghadimi, 2017).

Popularization of science is an attempt to depict scientific thinking so that everyone can understand the underlying science concepts and understand the essence of it. In other words, popularization of science is the attempt to generalize science, to spread scientific thought and culture, and to apply science at different levels of social life (Ghadimi, 2017).

Popularization of science, on the one hand, focuses on the public interest in science and technology. On the other, it deepens the public's understanding of science and technology. In other words, it deals with research in the three areas of society, science, technology, and government, and uses scientific methods for research in this field. Therefore, it is necessary to pay attention to the role of popularization of science in the light of the scientific and technological developments of the 20th century and the knowledge-based process of global developments in the 21st century (Ghadimi, 2017).

Popularization of science as an executive activity should have the characteristics that enable people to interact with science by incorporating science into the everyday lives of society so that society can enjoy the benefits of science. When science enters into the daily life of the population, arguably the social choices affect individuals. Increasing literacy rates and the number of graduates is an important factor, but not enough to achieve sustainable development. The type and quality of training are also crucial.

In the study of popular understanding of science, various imageries have been put forward to measure the popular understanding of science. Three aspects of this relationship are presented in Table 2.

Table 2: Paradigms, Problems and Proposals

Period	Attribution Problems	Proposals Research
Science Literacy 1960s onwards	Public deficit Knowledge	Literacy measures Education
Public Understanding After 1985	Public deficit Attitudes Education	Knowledge–attitude Attitude change Image marketing
Science and Society 1990s–present	Trust deficit Expert deficit Notions of public Crisis of confidence	Participation Deliberation "Angels" mediators Impact evaluation

Source: Bauer, Allum and Miller, 2007.

Popularization of science is to convey something to society so that we can receive it in the same way from society. That is, the effects of that science must be evident in society choices. Therefore, only a high literacy rate, which is a quantitative criterion, cannot reflect an increase in the level of social knowledge. For a country to benefit from the advantages of a knowledge-based economy, understanding of science and technology must penetrate all layers of society. The task of scientists is not only to write for each other but also society should be able to understand it. This can be done in the best form by popularization of science.

For example, popularization of science in America began with the democratization of society in the nineteenth century. The increasing economic importance of a skilled craftsman and mechanic in a modern economy led to a change in the medieval attitude towards workers. The new importance of craftsmen to the economy led to a major rethinking of the relationship between the craftsman and the scientist. Thus, advances in techniques and the material conditions of life were accompanied by corresponding changes in the perception and conceptualization of nature and society (Kuritz, 1981).

The link between work and science has strengthened the status and position of the craftsman or worker in many ways. The culture that emerged from the relationship in society not only increased the skills of workers and craftsmen but also elevated their status in society. Republican scientists like David Rittenhouse, Owen Biddle, and Charles W. Peale were all skilled artisans (Kuritz, 1981).

Today, the most prominent aspect of science is its ever-expanding development in a socially compatible way. In the past, only a handful of people were engaged in scientific activity and did so only on the sidelines of scientific institutions. However, as of today, scientific activity is an important and crucial part of social activity.

4. KNOWLEDGE-BASED ECONOMY INDICATOR AND POPULARIZATION OF SCIENCE

Popularization of science is one of the most influential parts of the structure of science-society communication. That is why science-society communication in general, and popularization of science, is important in both human and scientific development in national development programs.

Also, other aspects of development such as technology development, industrial development, economic development, and sustainable development are among the results and consequences of popularization of science. This makes the role of popularization of science more prominent in development. One of the pillars of national development is the development and improvement of human resources indicators. In terms of human development, there are three main goals for promoting science:

1. Convincing community members that employing a scientific approach is one of the best ways to deal with the various issues of everyday life—especially living in complex urban environments.
2. Empowering people to make the best use of technological goods and services that are diversified and rapidly expanding, and
3. Countering anti-science (superstition) and pseudoscience.

Today, in knowledge-based economies, the issue of literacy is not the challenge of education in the world. To deepen and institutionalize knowledge in the community, education must be inclusive at the highest levels among all segments of society, to provide for the expansion of knowledge and its application in the economic and social life of individuals. It may be argued that the education of the educated labor force is the first and most essential prerequisite for the introduction of knowledge into a country's economic system.

Following, we compare and analyze some of variables of Knowledge-Based Economy indicators introduced by the World Bank, according to popularization of science. In the pillar of ***education and human resources***, the three main variables that the World Bank has introduced are:

- Adult literacy rates, Figure 1;
- Secondary enrollment, Figure 2;
- Tertiary enrollment, Figure 3.

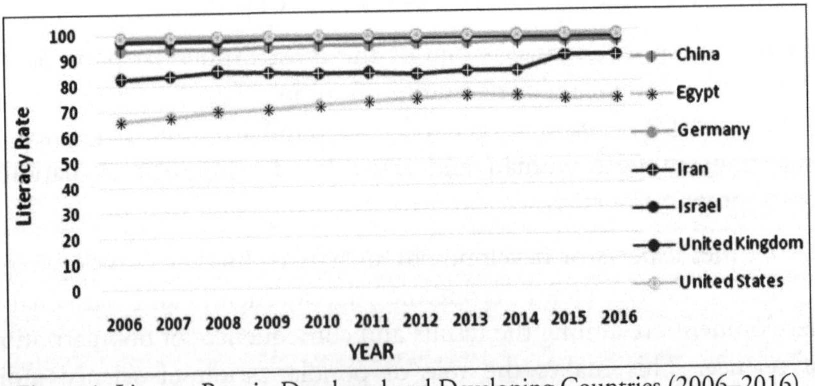

Figure 1: Literacy Rate in Developed and Developing Countries (2006–2016)
Source: World Bank data.

Indicators set by the World Bank to calculate the pillar of education and human resource are based on formal education calculations. While it ignored the tacit and productive knowledge that is integrated into a country's economic activities, the standard variables used as a measure of education and human resource are adult literacy rates, secondary enrollment, and tertiary enrollment. These indicators do not take into account the quality of education the students receive. The Education Index measures the amount of knowledge available to individuals regardless of the tacit knowledge accumulated in that community.

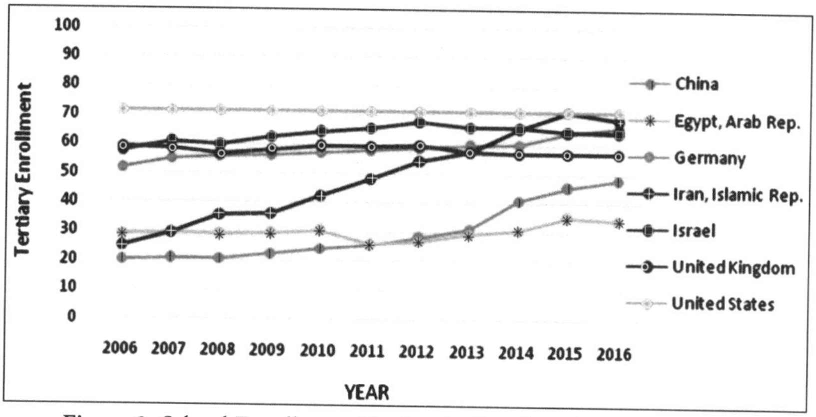

Figure 2: School Enrollment Tertiary in Developed and Developing Countries (2006–2016)

Source: World Bank data.

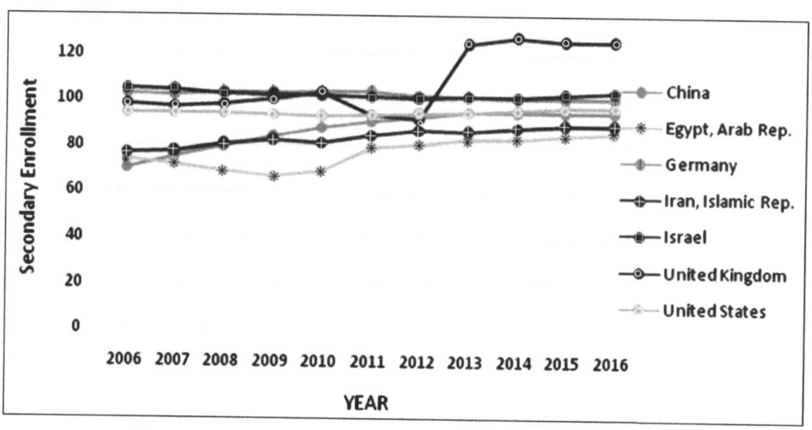

Figure 3: School Enrollment Secondary in Developed and Developing Countries (2006–2016)

Source: World Bank data.

Paying attention only to literacy rates and higher education enrollment may lead policymakers to the wrong idea of increasing the number of universities or schools. While the increase in universities and schools, regardless of popularization of science activities, leads to a widening gap between developed and developing countries.

In the pillar of ***Information infrastructure***, the two main variables that the World Bank has introduced are:

- Telephones per 1,000 persons (telephone mainlines + mobile phones), as shown in Figure 4;
- Internet users per 10,000 persons, Figure 5.

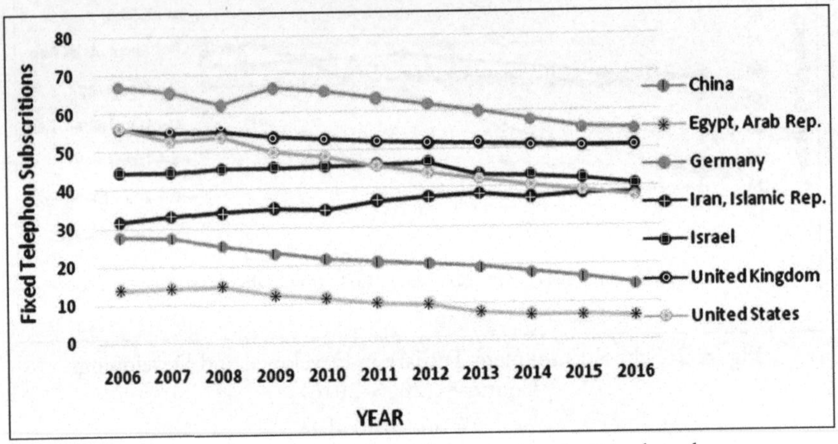

Figure 4: Fixed Telephone Subscriptions in Developed and Developing Countries (2006–2016)

Source: World Bank data.

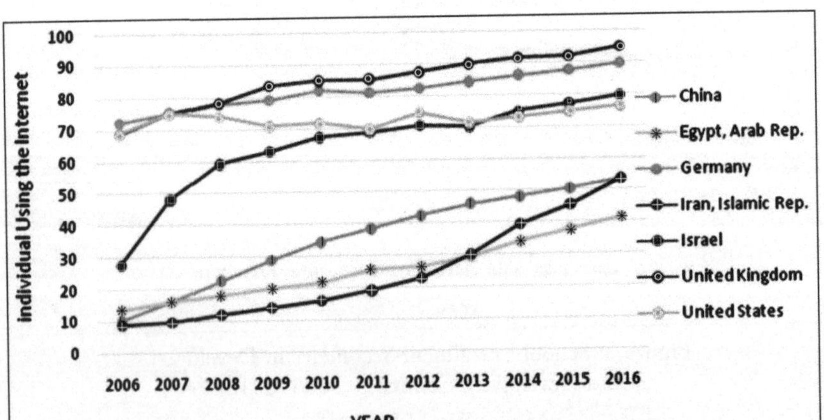

Figure 5: Individuals Using the Internet in Developing and Developed Countries (2006–2016)

Source: World Bank data.

One of the most obvious benefits associated with ICT usage is the increased flow of information and knowledge. But if this information and knowledge does not penetrate all layers of society properly, scientifically the increase in telephone and mobile subscribers will not increase public understanding in society.

In the pillar of ***Innovation system***, the main variables that the World Bank has introduced is "Scientific and technical journal articles, per million population".

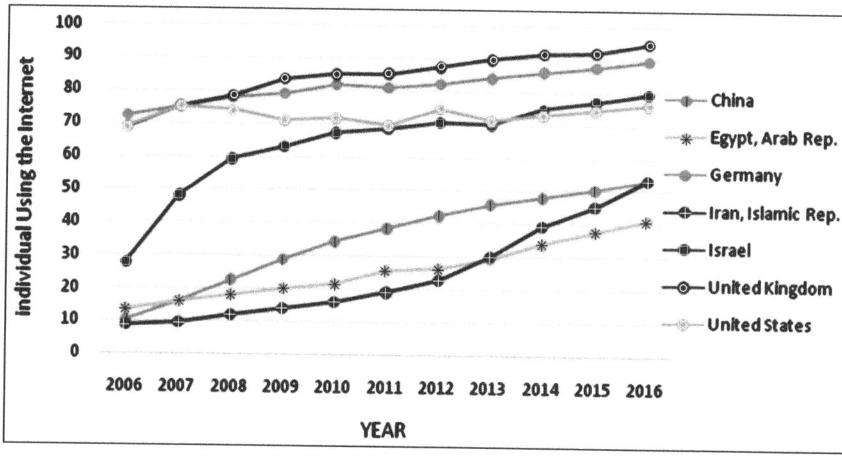

Figure 6: Scientific and Technical Journal Articles, Per Million Populations in Developed and Developing Countries (2006–2016)

Source: World Bankdata.

Also, increasing the number of scientific and technical articles in journals without adding to the public understanding will not have any impact on the development of the country.

These indicators are quantitative and there are fundamental differences in terms of the quality of these dimensions between developing and developed countries. In fact, the departure point of developing and developed countries is hugely different due to the existence of different institutions. Many studies have examined the impact of knowledge-based economy indicators on the economic growth and productivity of developing and developed countries. But just taking a look at the statistical data cannot make a good map for policymakers.

The tools and indicators for popularization of science can lead to the expansion and increase of the capacity for knowledge acceptance and innovation in society. Moreover, using these tools and indicators can be considered quantitative and qualitative dimensions of knowledge-based economy indicators together in each society. Because comparing countries only in the quantitative dimension can not provide us with a precise map and analysis.

5. EXPERIENCE OF SUCCESSFUL COUNTRIES IN PROMOTING SCIENCE

In 1993, the United Kingdom first spoke of a strategy to increase popular understanding of science and technology. The activities of popularization of science in England were very extensive and many of them were spontaneously carried out by schools, research, universities, associations, museums and so on. Indeed, there was some synergy between statesmen, scientists, and the public about the role of public understanding of science in boosting the British economy. Indeed, the necessity of scientific development and greater familiarity of citizens with science and research and development departments was a hallmark of most British science promotion programs. Before that, in 1972, the United States had cited research on popular science literature as one of its goals in engineering (Ghadimi, 2017).

In Germany, the promotion of more science is based on the link between the scientific and non-scientific members of society based on the mechanism of dialogue. The allocation of funds for popularization of science in German scientific policy is significant. In Germany, unlike the United Kingdom and the United States, popularization of science is closely linked to the production of science and the acquisition of new technologies and is not only used to increase the level of knowledge of ordinary people. China has enacted the Science and Technology Universalization Act, which helps facilitate public understanding, acceptance of science and technology, and participation in science. The law covers all publicizing activities by the government, community and public organizations, institutions, and local organizations. And citizen participation in public science and technology activities was counted as a right (Ghadimi, 2017).

6. CONCLUSION

To sum up, just taking a look at the statistical data cannot make a good map for policymakers. Consideration of the position of popularization of science in development programs is crucial because popularization of science plays an important role in achieving sustainable development, especially scientific development. To realize the scientific advancement of society, paying attention to the role of popularization of science in the development process of countries requires its development in macro programs and development of countries. This issue is scarce in the development plans and strategic documents of countries, which results in the slowdown of community development.

There is a positive correlation and interaction between "knowledge society" and "knowledge-based economy". Moving towards a knowledge-based economy requires raising citizens' interest in knowledge and resources and raising awareness because without these it would be difficult to produce modern knowledge.

REFERENCES

[1] Bauer, M.W., Allum, N. and Miller, S. (2007). What can we learn from 25 years of PUS survey research? Liberating and expanding the agenda. Public Understanding of Science, 16(1), 79–95.

[2] Chen, D.H. and Dahlman, C.J. (2005). The knowledge economy, the KAM methodology and World Bank operations. World Bank Institute Working Paper 37256.

[3] Ghadimi, A. (2017). Popularization of Science: Method, Indexes and Case Studies. Tehran: National Research Institute for Science Policy.

[4] Hausmann, R., Hidalgo, C.A., Bustos, S., Coscia, M., Simoes, A. and Yildirim, M.A. (2014). The Atlas of Economic Complexity: Mapping Paths to Prosperity. MIT Press.

[5] Kuritz, H. (1981). The popularization of science in nineteenth-century America. History of Education Quarterly, 21(3), 259–274.

[6] Stiglitz, J.E. (1998). Towards a new paradigm for development. Geneva: United Nations Conference on Trade and Development.

SECTION 4

STI Popularization Case Studies

Awakening the Nation to Genomics Technology: A Case Study of the Next Generation Biomedical Scientist (NGBS) Programme in Zimbabwe

M. Ndemera[1*], F. Muzenda[2], C. Bure[1], R. Karimanzira[1] and C. Masimirembwa[2]

[1]Government of Zimbabwe, Ministry of Higher and Tertiary Education, Science and Technology Development, P.O. Box CY7732, Causeway, Harare, Zimbabwe.

[2]African Institute of Biomedical Sciences and Technology, Block C, Wilkins Hospital, Corner Josiah Tongogara Ave and RekaiTangwena Ave, Harare, Zimbabwe

E-mail: *mkndemera@gmail.com

ABSTRACT: Zimbabwe has been plagued by economic woes for nearly two decades and this has negatively impacted the growth of the STI sector, which has experienced somewhat contracted growth. However, Zimbabwe has also recorded positive gains, with the expansion of the higher and tertiary education sector by the establishment of new technology-driven universities across the country. The Ministry of Higher and Tertiary Education, Science and Technology Development (MHTESTD) in collaboration with the African Institute of Biomedical Sciences and Technology (AiBST) developed a program to popularize genomics technologies nationwide. This program is called the Next Generation Biomedical Scientist (NGBS) Program. The NGBS program was aimed at creating awareness and capacity building in science, technology, engineering and mathematics (STEM) related disciplines, in this case, genomics. Another of the goals of the NGBS public engagement program was to challenge business and policymakers to consider investing in developing a research culture in learners at an earlier stage. Under the program, 375 learners (both urban and rural) from five provinces of Zimbabwe were trained in theoretic principles of forensic science, genomic medicine, and genomics technologies. Besides, the learners received lectures on biomedical entrepreneurship, scientific method development, and research proposal development, as well as an overview of careers in the biomedical sciences field. The trained learners were automatically entered into the NGBS competition and tasked to form teams and come up with community-based research projects in the areas of genomics, forensics and/or molecular diagnostics. 15 research projects were selected from the participating schools and funded. Because of the success of the NGBS training, and the potential to obtain innovative projects, the Government of Zimbabwe hosted an interface between the learners and stakeholders to popularize the various innovative learners' projects. In doing so, the Government of Zimbabwe successfully kick-started a process of knowledge sharing between the innovators (NGBS competition candidates) and those stakeholders that can commercialize their technological innovations.

Keywords: Genomics, Science and Technology, Science Education, Zimbabwe, Capacity Building.

1. INTRODUCTION

Zimbabwe has a clear-cut Science, Technology and Innovation policy. Zimbabwe's overall Science, Technology and Innovation (STI) regulatory framework is outlined in a number of key policy documents that include the Research Act of 1986 and its subsequent amendments, the Science and Technology (S&T) Policy of 2002, the Biotechnology Policy of 2005, the Biotechnology Act of 2006 and the ICT Policy Framework of 2006 and the current Science, Technology and Innovation (STI) Policy of 2012. Two of the goals of the STI policy include;

1. Searching for scientific solutions to global environmental challenges.
2. Mobilize resources and popularize science and technology.

To realize the above goals, it is commonplace that a significant investment is necessary to kick start and sustain policy gains outlined above. However, Zimbabwe has been plagued by economic woes for nearly two decades and this has negatively impacted the growth of the STI sector, which has experienced somewhat contracted growth. Despite this fact, Zimbabwe has also recorded positive gains, with the expansion of the higher and tertiary education sector by the establishment of new technology-driven universities across the country. To date, the Government of Zimbabwe has invested in 6 innovation hubs in strategically placed universities nationwide to foster STI development as a means of achieving the goals outlined in the STI policy. The innovation hubs are envisaged as STI focal points, whereby active knowledge transfer between researchers and business experts, on the one hand, and industry, government and representatives of academia, on the other hand, can take place. Here, decision-makers can meet and brainstorm with scientists and business experts and discuss their complex business challenges.

Several key programs have been developed by the Government of Zimbabwe towards the fulfillment of strategic goals of the STI policy outlined above. The Ministry of Higher and Tertiary Education, Science and Technology Development (MHTESTD) is responsible for "Program 4: Application of Science and Technology to Enhance Health and Justice Delivery Systems". Under this program, genomics technologies have been identified as an effective solution to the enhancement of health and justice delivery systems in Zimbabwe. Under the Department of Projects and Technology Transfer within the Ministry, a program was developed to popularize genomics technologies and in so doing, identify and promote key technologies for integration into the various innovation hubs. This program is called the Next Generation Biomedical Scientist (NGBS) Program.

The NGBS program was aimed at creating awareness of a science, technology, engineering and mathematics (STEM) related discipline, in this case, genomics. The Government of Zimbabwe recognizes under Program 4, that training is a critical component of human capacity building and that through the NGBS program capturing talent from young minds and in so doing fostering an STI research culture from a tender age is important. As such, the Government of Zimbabwe has fully supported the NGBS program as a means of popularizing STI at the national level. Therefore, this paper seeks to outline the impact and success of the NGBS program in popularizing science and technology in Zimbabwe.

2. METHODOLOGY

Initially, a comparative analysis of the state of science and technology appreciation and adoption was carried out. This analysis involved probing the instruments and documentation around popularization of S&T using a funnel approach, beginning at the inter-continental level and narrowing down to Zimbabwe, an example of a developing nation trying to harness its S&T potential. This was followed by describing in detail, the Next Generation Biomedical Scientist (NGBS) model, taking into consideration the reach and impact of the model in the popularization of S&T in Zimbabwe. Finally, a synopsis of the role of the NGBS model and program in popularizing science and driving the industrialization agenda was explored to devise recommendations for the furthering of S&T research and innovation in Zimbabwe.

3. RESULTS AND DISCUSSION

3.1 The Status of Science, Technology and Innovation (STI) in Africa

The growth and pace of Science, Technology, and Innovation (STI) have been described as slow compared to STI growth in the 1st world nations (Godwill, 2014). The African Union Science, Technology and Innovation Strategy for Africa 2024 (STISA-2024) recognize that science, technology, and innovation are multi-function tools and enablers for achieving continental development goals (African Union Commission, Agenda 2063: The Africa We Want THE VOICES OF THE AFRICAN PEOPLE. Final Edition, 2015). In this strategy, infrastructure development including building and/or upgrading research infrastructure, technical competencies, innovation & entrepreneurship as well as an enabling environment for STI development have been identified as pillars in the realization of a S&T driven continent (African Union Commission, 2014). However, human capacity gaps remain a limiting factor to the realization of the objectives of the STISA 2024 with very limited STI skill levels and competencies being harnessed at both the continental and national levels by member states. Further, the situation is worsened by the constant flight and brain drain being experienced as many a scientist leaves for the 1st world in search of greener pastures. The STISA 2024 recognizes the need to improve STI status through human capacity development (African Union Commission, 2014). However, there needs to be a mechanism of retaining this human capacity to reap the benefits at the continental and member state level.

In addition to limited human capacity, inadequate STI policies have greatly hampered STI in Africa. It was noted that though member states are seized with developing STI policies, many of the officials involved in or responsible for drafting policy documents do not have the necessary skills or training and have no experience in evidence-based policy making and therefore very limited evidence-based policy development is occurring in Africa (African Union Commission, 2014). This is a disadvantage as it exposes member states to expending time and resources developing polices and guidelines that are not fit-for-purpose. As such, the implementation of these policies is expected to yield no tangible results.

Even though research and development are not fully utilized as a tool for development, it does not mean that there aren't any meaningful research taking place in Africa and Zimbabwe specifically. The need to ensure public understanding and utilization of science through exploring how to deliver scientific information to laypeople in a manner that makes it attractive and enjoyable has been identified as a key driver to the adoption of science in Africa (United Nations Economic and Social Council, 2003). However, it is understandable that the majority of research and innovation is carried out by individuals who may not have access to infrastructure and technology incubation and transfer facilities. The reasons vary but chief among them is lack of effective identification and communication strategies for research and innovation. Communication of research outcomes is an integral part of the research, innovation, and technology transfer process. Table 1 shows the current status of S&T documentation in Zimbabwe according to the findings of the Zimbabwe National Research and Development (R&D) Survey (2017). The Zimbabwe National Research and Development (R&D) Survey is an annual survey carried out by the Ministry of Higher and Tertiary Education, Science and Technology Development to obtain information on the status, trends and statistics of national R&D activities that are conducted for two sectors namely institutions of higher learning and research institutes.

Effective research communication enables the establishment of necessary linkages and structures that will translate an idea into a product or service ready for commercialization. This includes improving on the governance systems and policies which can scientific and

Table 1: Current Status of Published S&T Documentation in (Government of Zimbabwe, 2018)

Articles	*2011*	*2012*	*2013*	*2014*	*2015*	*2016*
Natural Sciences	7	41	63	89	112	131
Engineering & Technology	0	20	34	59	55	63
Medical & Health Sciences	55	90	76	43	59	104
Agricultural Sciences	33	104	150	144	180	181
Social Sciences	45	110	119	103	159	137
Humanities	142	286	352	456	520	609
Books	*2011*	*2012*	*2013*	*2014*	*2015*	*2016*
Natural Sciences	0	1	0	0	3	12
Engineering & Technology	0	2	1	0	10	4
Medical & Health Sciences	0	0	0	0	0	0
Agricultural Sciences	4	3	8	6	6	6
Social Sciences	1	3	2	3	6	11
Humanities	7	9	9	13	13	19
Conference proceedings	*2011*	*2012*	*2013*	*2014*	*2015*	*2016*
Natural Sciences	4	15	20	19	28	30
Engineering & Technology	0	0	14	19	58	59
Medical & Health Sciences	0	0	4	10	7	3
Agricultural Sciences	19	44	44	61	60	60
Social Sciences	8	10	14	17	35	64
Humanities	148	188	172	194	223	279
Audios	*2011*	*2012*	*2013*	*2014*	*2015*	*2016*
Natural Sciences	0	0	0	0	0	0
Engineering & Technology	0	0	0	0	0	2
Medical & Health Sciences	0	0	0	0	0	0
Agricultural Sciences	0	0	1	0	0	0
Social Sciences	0	0	0	0	0	1
Humanities	0	0	0	0	0	3
Documentaries	*2011*	*2012*	*2013*	*2014*	*2015*	*2016*
Natural Sciences	0	0	0	0	0	0
Engineering & Technology	0	0	0	0	1	1
Medical & Health Sciences	0	0	0	0	0	0
Agricultural Sciences	0	0	0	0	0	0
Social Sciences	0	0	0	0	0	1
Humanities	0	0	0	0	0	0

technological processes including research and scientific projects (Godwill, 2014). Besides, there is a need to increase documentation of scientific outputs through publishing acquired knowledge and empirical research, for example in peer-reviewed journals. To this effect, the need to train scientists on how to communicate their messages effectively cannot be understated (United Nations Economic and Social Council, 2003). In Zimbabwe, there is a steady increase in the number of scientific publications over the past years. However, more needs to be done to increase the number of high-quality publications across all sectors.

3.2 NGBS Model

keeping with the cabinet approved program on the application of science and technology to enhance health delivery systems (program 4 in short), the NGBS training program was initiated as one of the public engagement initiatives under program 4 – "*Application of science and technology to enhance health delivery systems*". The Government of Zimbabwe through its Ministry of Higher and Tertiary Education, Science and Technology Development (MHTESTD), in collaboration with the African Institute of Biomedical Sciences and Technology (AiBST), in a typical public-private partnership, has realized an opportune moment to advance its technology transfer mandate by fostering necessary linkages between researchers/innovators and businesses. One of the goals of the NGBS public engagement program was to challenge business and policymakers to consider investing in developing research culture in learners at an earlier stage. One of the aims was to build public interest in the power of scientific and technological innovations in resolving health and security challenges at the grassroots level and ultimately to develop an appreciation by ordinary citizens, business and policymakers of the potential within Zimbabwe to generate solutions for various health-related societal challenges by harnessing the brains of young people. For this process to be relevant and sustainable, NGBS program uses a model of engagement which is inclusive of key stakeholders in health innovation and development, the students themselves, parents, community leaders, business/industry players, teachers and national science think-tanks. This fits into the STISA-2024 strategy, which was crafted in response to the demand for science, technology, and innovation to impact critical sectors including agriculture, energy, environment, health, infrastructure, mining, security and water among others (African Union Commission, 2014). Further, it is recognized in the STISA-2024 strategy that communication and outreach on STI are important to secure necessary political buy-in at government level, raise wider public awareness at the grassroots level, and garner support from all key stakeholder groups. Therefore, the main objectives of the NGBS program were to;

1. To encourage and develop a culture of research and innovation in the young generation particularly in biomedical sciences.
2. To develop a culture of applying the critical thinking methodology towards problem-solving in individuals from a young age (secondary school level).

The NGBS program was a nationwide program targeting marginalized secondary school pupils to raise awareness of genomics technologies and careers in genomics technology. The main thematic areas were projects centered around the use of genomics for health and forensic research and innovation. The NGBS training were conducted in five out of Zimbabwe's ten provinces namely Manicaland, Bulawayo, Masvingo, Matebeleland South, and Midlands. The target group for the program was secondary level learners, particularly those in Form three and Lower Sixth Form studying science, technology, engineering and mathematics (STEM) related subjects. Under the program, 375 learners from the five provinces were trained in theoretic principles of forensic science, genomic medicine, and genomics technologies. Also, the learners received lectures on biomedical entrepreneurship, scientific method development, and research proposal development, as well as an overview of careers in the biomedical sciences field. The learners also benefitted from practical training in laboratory techniques (standard operating procedures, sample preparation, etc.) and genomics techniques (DNA amplification, restriction fragment length polymorphism (RLFP), gel electrophoresis, etc). Of the 375 learners reached, 53% were female. The learners were drawn from both rural and urban schools.

The fact that the NGBS program had reached as far as rural schools is an important milestone, given that rural schools have been known to be marginalized and lagging concerning science, technology and innovation programs. However, this does not in any way reflect the inability concerning science, technology, and innovation. It was discovered that learners in rural schools are very intelligible in science subjects but just lack the opportunity to express their capabilities which is a barrier to realizing their potential. On a national scale, this is a disadvantage as a lot of innovative ideas can be missed. The NGBS program sought to unearth such hidden talents by providing an equal opportunity to both rural and urban learners to be trained and enhance capacity in genomics technologies. The first outcome of the NGBS program was to raise students' awareness of research culture, develop basic research skills, and gain experience in problem-solving projects. Learners enhanced their capabilities to carry out further university-level research projects through a rare opportunity to conduct real-life research thereby being inspired to pursue research as a career.

The trained learners were automatically entered into the NGBS competition and tasked to form teams and come up with community-based research projects in the areas of genomics, forensics and/or molecular diagnostics. The submitted projects were supposed to explore the application of genomics technologies to solve community problems and were supposed to involve community leadership structures, scientist-mentors from local universities and local health experts. In so doing, the learners practiced engaging policymakers in pushing research-driven agendas for development and to engage their communities in co-creating solutions for local health problems. 15 research projects were selected from the participating schools and funded. Figure 1 is an overview of the NGBS public engagement program.

The Government of Zimbabwe realizes the need to build public interest in the power of scientific and technological innovations in resolving health and security challenges at grassroots level and ultimately to develop an appreciation by ordinary citizens, businesses and policymakers of the potential within Zimbabwe to generate solutions for various health-related societal challenges. Public appreciation of STI is critical for successful implementation of STI policies and programs at national, regional and continental levels and one awareness method involves communicating the practical and tangible benefits of STI must be communicated to key stakeholders (African Union Commission, 2014). This can only be achieved by harnessing the brains of young people, at first instance. Given the success of the NGBS training and the potential to obtain innovative projects the Government of Zimbabwe hosted an interface between the learners and stakeholders to popularize the various innovative learners' projects. Stakeholders were drawn nationwide, from other government agencies, research institutions, academic institutions and the business sector at large. The main objective of the stakeholder interface was to identify potential R&D and innovation projects that can be further refined and/or scaled up as well as to popularize science, technology and innovation efforts in Zimbabwe. Using this NGBS program public engagement initiative, the Ministry of Higher and Tertiary Education was able to fulfill one of its core mandates of technology transfer by kick-starting relationships between the innovators (NGBS competition candidates) and those stakeholders that can commercialize their technological innovations.

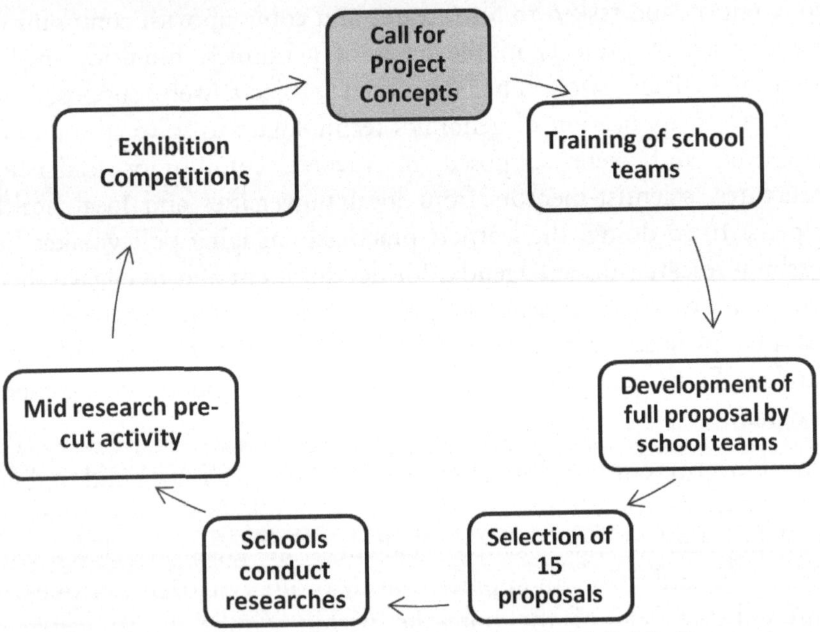

Figure 1: Next Generation Biomedical Scientist (NGBS) Public Engagement Program; April 2018–June 2019

3.3 Future Outlooks: Towards Comprehensive Policies and Technology Transfer

The mandate of the Ministry of Higher and Tertiary Education Science and Technology Development (MHTESTD) is to train and develop skilled human personnel and to promote the development and application of science and technology in all sectors of the economy in Zimbabwe to drive the industrialization and modernization agenda. Currently, Zimbabwe has, on average less than 15% skill levels in Science, Technology, Engineering and Mathematics (STEM) disciplines (engineering, agriculture, health, etc.). Furthermore, national universities have remained focused on teaching, research and community services with little or no emphasis on applied research to support the industry as evidenced by the 2017 Africa Science, Technology and Innovation Indicators (ASTII) report (Gouvernement of Zimbabwe, 2018). It is for this reason that the Ministry of Higher and Tertiary Education, Science and Technology Development introduced two more pillars to higher education—innovation and industrialization—critical elements necessary to grow the economy by the desired date of 2030. Thus, the continuum from teaching and learning to the application of knowledge to develop the economy—Education 5.0—required the necessary infrastructure to ensure effective translation of this knowledge into goods and services. It is for this reason that the Ministry embarked on a program to establish Science Parks/Innovation Hubs (SP/IH) at all tertiary education institutions as a standard for higher learning—both public and private. Figure 2 is an attempt to illustrate the interlinking of SP/IH within the framework of Education 5.0.

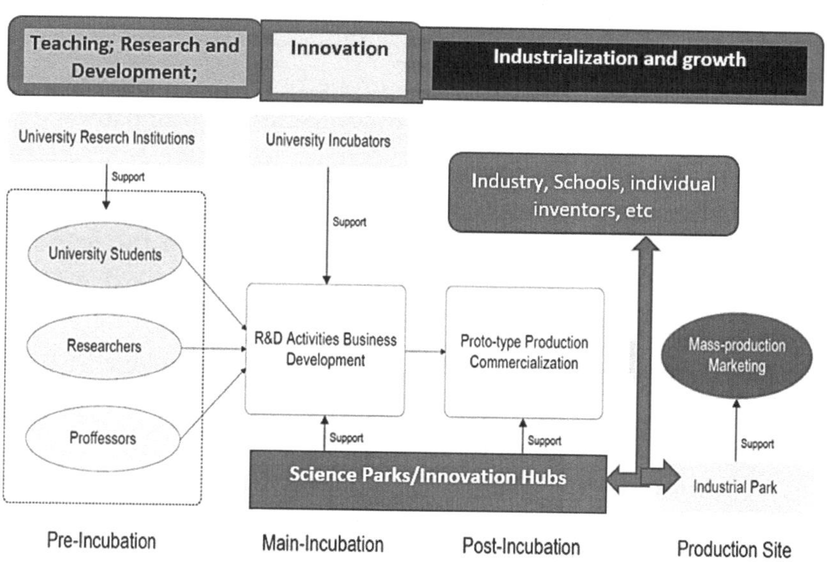

Figure 2: Education 5.0 and Role of SP/IH

SP/IH should be viewed as entities that can support business creation promotion through the identification of potential entrepreneurship projects. Therefore, it is imperative to inculcate an entrepreneurship culture in the custodians of the SP/IH and to actively enhance the entrepreneurial and management skills of all the direct stakeholders of the SP/IH. Success factors should be inherent in a SP/IH include a robust national system of innovation as well as dedicated start-up support services to offer would-be entrepreneurs and those in the start-up process entrepreneurial support, information and necessary linkages e.g. access to a portfolio of investors/supporters/stakeholders. These should be backed by formalized industry/academic policy/legal frameworks including incentives to facilitate and promote domestic innovativeness, transfer of technologies and know-how and creation of start-up ventures and access to appropriately skilled human capital for technical support, mentorship and supervision.

4. CONCLUSIONS

The NGBS model proved to be a unique and fit for purpose initiative with the potential of being scaled up. The NGBS program is a useful avenue for the popularization of science and technology and a source of untapped research and innovation. It is important to consider significant investment in programs such as the NGBS program as a means to achieve and sustain the objectives outlined in Zimbabwe's Science, Technology and Innovation policy. Moreover, there is a need to develop and appropriate intellectual property (IP) policy to safeguard the gains of various research and innovation efforts in Zimbabwe.

5. ACKNOWLEDGMENTS

The authors would like to acknowledge support from the following entities:

- Wellcome Trust, UK.
- African Institute of Biomedical Sciences and Technology (AiBST), Zimbabwe.
- Government of Zimbabwe:
  - Ministry of Higher and Tertiary Education, Science and Technology Development.
  - Ministry of Primary and Secondary Education.

REFERENCES

[1] African Union Commission (2015). Agenda 2063: The Africa We Want THE VOICES OF THE AFRICAN PEOPLE. Final Edition.

[2] African Union Commission. (2014). Science, technology and innovation strategy for Africa 2024. From https://au.int/sites/default/files/news events/workingdocuments/33178-wd-stisa-english_-_final.pdf

[3] Godwill, E.A. (2014). Science and Technology in Africa: The key Elements and Measures for Sustainable Development. Global Journal of Science Frontier Research: G Bio-Tech & Genetics.

[4] Government of Zimbabwe (2018). Zimbabwe National Research and Development Survey. Harare: Zimbabwe.

[5] United Nations Economic and Social Council (2003). Making science and tecnology work for the poor and for sustainable development in africa. Addis Ababa, Ethiopia.

ABBREVIATIONS

AiBST	African Institute of Biomedical Sciences and Technology
ASTII	Africa Science and Technology Innovation Indicators
AU	African Union
DNA	Deoxyribonucleic Acid
MHTESTD	Ministry of Higher and Tertiary Education, Science and Technology Development
NGBS	Next Generation Biomedical Scientist
R&D	Research and Development
RFLP	Restriction Fragment Length Polymorphism
S&T	Science and Technology
SP/IH	Science Park/Innovation Hub
STEM	Science, Technology, Engineering and Mathematics
STI	Science, Technology and Innovation
STISA	Science, Technology and Innovation Strategy for Africa

Policy Making for Popularization of Traditional Medicinal Knowledge

Gloria Susan Cherian
Research Associate, NAM S&T Centre, India Habitat Centre,
Lodhi Road, New Delhi, India
E-mail: gloriacherian27@gmail.com

ABSTRACT: Throughout history, substantial contributions have been made in the field of traditional medicine by the indigenous people, their cultural perspectives and through their approach towards medicinal practices. Traditional healers and their natural remedies utilizing medicinal plants are known to play a key role in the health of millions of people. However, indigenous knowledge using plant resources have been constantly diminishing due to the changing perception of local people, the increasing influence of global commercialization and the socio-economic transformation.

The main reason for this decline can be directly linked to less shared knowledge on folk medicines as with the advancements in modern science and medicines. Further, rare documented evidence of less exploited herbs or less accessible information on literature and knowledge regarding traditional medicine also confirms the consensus on gradual extinction of traditional medicinal knowledge (TMK). This stirs up a need for formulating and introducing techniques to popularize the indigenous knowledge about traditional medicine and its enculturation in the national STI policy of all developing countries, for primary health care needs. It is important to note that the herbal medicines being less concentrated and less toxic are used in much lower doses than the orthodox medicine.

With the inclusion of two case studies, this research aims to promote traditional medicine as a vehicle for improving the lifestyle of indigenous people in developing countries and to encourage utilization of indigenous knowledge as a source of income generation having tremendous entrepreneurial potential. The paper also focuses on the need for policy making and its implementation, to protect and popularize TMK.

Keywords: Indigenous Knowledge, Traditional Medicine, Lifestyle, Income Generation, Entrepreneurial Potential.

1. INTRODUCTION

Indigenous knowledge (IK) encompasses wide ranging subject areas from art to agriculture, including medicinal uses of plants and the traditional systems of medical diagnosis. It by and large incorporates the local knowledge embodying traditional lifestyles, and the acquired or learned skills that lead the practitioners to new discoveries and advancements. According to Ryan Abbott (2014), IK exists in indigenous or local communities as the intellectual and intangible cultural heritage, practices and knowledge systems that are passed down over generations; or might be archived in freely accessible, composed and electronic media.

Whereas, traditional medicinal knowledge (TMK) refers specifically to the knowledge of indigenous plants and their healing properties; the approaches, experiences and beliefs that are used in the maintenance of health; in addition to the spiritual therapies, manual techniques and exercises which may be applied singularly or in combinations so as to treat, diagnose and prevent physical & mental illnesses or maintain well being of an individual. Traditional medicine (TM) in many cultures functions as a comprehensive system of health care refined over thousands of years and is also referred to as indigenous or folk medicine. Some of the best-known TM systems including the traditional Indian medicine (Ayurveda), traditional Chinese medicine, Iranian traditional medicine, traditional African medicine and traditional Perso-Arabic (Unani) medicine have safeguarded the health of indigenous communities for generations as stated by Anyaoku E.N. *et al.* (2015). However, there is a growing consensus on the gradual extincttion of IK including TMK and the consequences would include the vital knowledge gradually being lost, that might contribute to the survival struggle for man and animals in the near future.

TMK is rapidly disappearing, owing to cultural change and declining access to sources of natural medicinal products, in both urban and rural areas. Most villages in the world are no longer surrounded by the natural habitat that formerly served as a medicine cupboard, and the bodies of folk knowledge that have accumulated and been honed for thousands of years are disappearing at an alarming rate. In some cases this loss may actually confer net health benefits; but modern society will never know what effective medicinal treatments are being lost (Daily G.C. and Ehrlich P.R., 1996).

TMK has historically been at odds with modern intellectual property (IP) systems designed to protect innovations such as new pharmaceutical drugs. However, as the financial value of many forms of traditional medicine becomes recognized, traditional knowledge holders and nations rich in genetic resources are arguing for greater protection through non-conventional systems of intellectual property protection. Traditional knowledge holders are increasingly demanding fair and equitable distribution of benefits from the commercialization of traditional medicine, as well as the prior informed consent of indigenous peoples to prevent misappropriation.

This paper primarily aims to popularize TMK using illustrative case studies, in order to help traditional knowledge holders make informed decisions about the best use of their knowledge for income generation, entrepreneurship, and in obtaining a better lifestyle; and also to encourage practitioners, government representatives and third-party collaborators to deliberate upon implementation of effective policies and IP issues related to TMK.

2. IMPORTANCE OF TMK

WHO has acknowledged that "traditional as well as complementary and alternative medicine (CAM) have many positive features, and that TM and its practitioners play an important role in treating chronic illnesses and in improving the quality of life of those suffering from minor illnesses or even certain incurable diseases" (Fifty Sixth World Health Assembly, 2003). Traditional knowledge about the medicinal properties of plants has been instrumental in pharmacological development. For example, the active ingredient in pain relievers, Aspirin (acetyl salicylic acid) was first discovered by indigenous people who utilized the bark of the willow tree. Moreover, indigenous people with their decades of personal experience and the ecological relationships between them, to a great extent complement this comprehensive system of healthcare. For instance, in North America, the indigenous people helped newcomers cure the life-threatening disease scurvy through tonics made from conifer needles that were rich in vitamin C. (Jesse Popp, 2018).

Additionally, TM when compared to conventional medicine is more affordable, strengthens the immune system, has fewer side effects, is easier to obtain than prescription medicine, facilitates natural healing while maximizing the body's inherent healing ability, and also stabilizes hormones and the body's metabolic rate. The major differences in TM and modern medicine (MM) as pointed out by Andrea R. and Priya S., 2015 are briefly listed in Table 1.

Table 1: Major Differences between TM and MM

	Traditional Medicine (TM)	*Modern Medicine (MM)*
Knowledge protection	Open access but social or legal restrictions may govern who can use certain knowledge, under what conditions and with what benefit for knowledge holders	Closed, patent-protected
Formulation	*Ad hoc* during consultation with the patient	Predetermined, once approved in clinical trials, formulas cannot be changed unless retested
Regulation	Usually loose, sometimes restricted for use or dissemination; rules and standardization may vary	Extremely tight
Testing	No formal testing; understanding of effectiveness is handed down through generations	Rigorous trials in different phases (first testing for safety, then efficacy) means bringing a drug to market costs billions of dollars
Dosage	Unfixed; usually less but the amount of active ingredient can vary hugely	Standardized medicines given in fixed doses that vary with age or weight, or disease severity
Consultation	Lengthy, and the patient is asked a wider range of questions than just about their symptoms	Consultations in both primary and secondary care tend to be brief and focused, especially as national health systems come under financial strain
Training	Lengthy training; knowledge often passed one-to-one through families; practitioners are often born into a family of healers	Lengthy and often vocational: health professionals go through formal training in schools and universities

As opposed to the relatively modern medicinal practices, TM has the benefit of substantial prior clinical use as well as stronger cultural associations. This can provide evidence of safety and efficacy and could thus result in being more readily accepted by some populations (Directive 2004/24/EC).

2.1 TMK for Raising Entrepreneurial Potential

Entrepreneurship has a tremendous effect in empowering and transforming the society. Increasing socio-economic awareness, need for additional income, utilization of extra time, constant motivation by the government institutions and the impact of role models are some of the factors responsible for the development of entrepreneurship in the developing countries that can also support communities to expand their business both in herbal medicinal product diversification and also in gaining access to international markets. Success in business makes entrepreneurs financially independent and also gives them the strength to overcome setbacks in life.

Indigenous people often possess knowledge related to harvesting, preparing herbs, and about the medicinal use of plants. This information can be invaluable not only to communities who have historically used herbal medicines, but also for export and use outside of its traditional environment. In other words, TMK can enable ethnic minority groups to become entrepreneurs in providing herbal medicine for health care of the society. Entrepreneurs may not just provide raw materials for pharmaceutical companies, but holders of TMK often have valuable knowledge for new drug development. It is estimated that for every 10,000 pure compounds that are biologically evaluated, only one achieves regulatory approval as studied by P.R. Vagelos in 1991. Even though new drug development is an expensive and risky venture, pharmaceutical companies invest billions of dollars annually in the hope of developing new chemical entities that are safe and effective, and which can be manufactured in a cost effective way.

While some medicinal plants are cultivated commercially, most continue to be collected from the wild. Pre-industrial communities have been responsible for the discovery of most of the medicinal plants in use today, and many communities are still involved in the wild collection, domestication, cultivation and management of medicinal plant resources. This economic activity supports many indigenous peoples and local communities, a benefit that in turn provides incentives for the conservation of TM.

The search for plant and animal species from which medicinal drugs and other commercially valuable compounds can be obtained is called bio-prospecting, and can be beneficial to indigenous communities in the developing countries. B. Barrett *et al.*, 1999 stated that as much as one-third to one-half of pharmaceutical drugs are originally derived from plants. Some prominent examples include digitalis, a popular cardiac medication identified as the active component of the foxglove leaf; morphine and codeine which alleviate pain, derived from cultivated opium poppy; atropine for disorders involving the autonomic nervous system, from the nightshade plant; the anti-cancer drug Taxol derived from the bark of the Pacific yew tree, and Aspirin from willow bark (Carmen and Carlos, 2008).

TMK thus provides a valuable guidance in selecting and obtaining plant material of potential therapeutic interest. However, if un-regulated, bioprospecting can result in over-exploitation of limited resources. Habitats that support medicinal plants are being destroyed by over-harvesting and new commercial developments. Many traditional medicines now face extinction with serious consequences for local communities, for example, licorice root, the most commonly used Chinese herb is now threatened (Dan Bensky *et al.,* 2014) and is now grown in less than half its previous area as a result of excessive harvesting and habitat destruction.

The widespread use of TM can actually result in traditional health care becoming a lucrative, multinational business. For example, in India, an outlay of 298 million USD has been earmarked for the Ayurveda, Yoga and Naturopathy, Unani, Siddha and Homoeopathy (AYUSH, see appendix) sector in the financial year 2019.

Further, as the financial value of TM gets recognized, TK holders and nations rich in genetic resources should also aim for greater protection through non-conventional systems of IP protection.This will therefore help the ethnic minority groups to become herbal medicine entre-preneurs.

2.2 TMK for Income Generation

The striking coincidence between the uses of indigenous medicinal plants and scientifically-proven pharmacological properties, shows that the traditional remedies are an important and effective part of indigenous healthcare systems which are totally dependent on traditional healers. More detailed phytochemical and pharmacological studies of traditionally-used medicinal plants are thus an important line of research to be pursued.

It is understood that medicinal plants provide huge opportunities for development of local communities and for their livelihood improvement. However, indigenous people are often deprived of the benefits from these resources whenever they are commercialized. Proper management of high-value and high-priority medicinal plants could serve as a sustainable income source for the communities which would in turn help generate incentives for biodiversity conservation, thus ensuring the long-term availability of medicinal plants for indigenous and commercial uses.

The need of the hour therefore remains to train local communities for proper exploitation and preservation of plants with high values; for initial processing, marketing and selling of medicinal plants; and for help in producing sophisticated products like instant powder tea or syrup, etc. It also involves protecting innovations of new formulations and documenting it, as its financial value becomes recognized and lastly, the equitable distribution of benefits to the TK holders from the commercialization of TM.

If natural medicinal plants are exploited in a sustainable way and cultivated more for herbal medicine industry, it would definitely contribute towards the country's economic growth. It can therefore be said that TM is not only a vital source of health care, but also an important source of income for many communities.

2.3 TMK for a Better Lifestyle

TM practices as stated by Ryan B. Abott (2011), tend to be characterized by a holistic and highly individualized approach to treatment, with an emphasis on maximizing the body's inherent healing ability. It involves patients as active participants in their own treatment, and addresses physical, mental, and spiritual attributes of a disease, thereby placing a strong emphasis on prevention and wellness. Extracts from plants have been found to be useful for treating diseases such as malaria, diabetics, epilepsy, dementia, sickle cell disorders and inflammation (Adeleja, 2006). Some commonly used medicinal plants are listed along with their uses in the Table 2.

Table 2: List of Some Medicinal Plants and their Uses

Plant	*Common Name*	*Botanical Name*	*Part/s Used*	*Uses*
	Lemon grass	*Cymbopogon citratus*	Leaves	Alleviates laryngitis and sore throats; has anti-pyretic property which reduces high fevers; treats abdominal pain, headaches, joint pains, muscle pains, digestive tract spasms, muscle cramps, stomachache; inhibits bacteria and yeast growth; and is beneficial for nervous and stress-related conditions.
	Indian lilac/Neem	*Azardirachtaindica*	Leaves, Rhizome	Heals minor wounds, eczema, epilepsy & hypertension; used as a sedative and analgesic
	Indian Pennywort/ Gotu kola	*Centellaasiatica*	Leaves	Treats ulcers, leprosy, skin injuries, decreases capillary fragility and stimulate lipids and protein necessary for healthy skin, revitalizes the brain and nervous system

Plant	*Common Name*	*Botanical Name*	*Part/s Used*	*Uses*
	Ginger	*Zingiber officinale*	Rhizome, leaves	Reduces nausea, pain, blood pressure and inflammation; eases headaches and flu
	Coriander	*Coriandrum sativum*	Leaves, dried seeds	Treats upset stomach, loss of appetite, hernia, nausea, diarrhea, bowel spasms, intestinal gas, measles, hemorrhoids, toothaches, worms, and joint pain, as well as infections caused by bacteria & fungus
	Winter savory	*Saturejamontana*	Leaves	Antiseptic, treats bug bites and other wounds, itching & can help sooth a sore throat or ease indigestion
	Grapeseed oil	*Vitis vinifera*	Grape seed extract and grape peel extracts	Very good source of Vitamin-E; works as a fat-soluble antioxidant that protects cells from damaging free radicals associated with cancer & heart disease; lowers LDL (bad) cholesterol and BP

Plant	*Common Name*	*Botanical Name*	*Part/s Used*	*Uses*
	Gooseberry	*Emblica officinalis*	Fruit	Rich in Vitamin-C; cures cough and cold, hyper acidity &diabetes; used as Laxative
	Fenugreek	*Trigonella foenum-graecum*	Leaves, dried seeds	Encourages weight gain, inhibits liver cancer, lowers blood cholesterol levels, treats inflammation and stomach ulcers, painful menstruation & labor pains; drain off sweat ducts
	Aloe vera	*Aloe barbadensis miller*	Gel, latex	Treats wounds, hair loss, hemorrhoids, skin problems and digestive issues

Bioactive compounds derived from currently used herbal medicines are more likely to have minimal toxicity, and a long history of clinical use suggests that a herbal medicine may be clinically effective. Plant-derived compounds used as drugs are generally used in ways that correlate directly with their traditional uses as plant medicines (Daniel and Norman, 2001). This gives an insight into how useful TMK is in helping to improve an individual's health condition. It is therefore imperative to protect this knowledge from uses that may erode it or negatively affect the life or culture of the communities that have developed and applied it, as it also has a direct positive role in creating sustainable livelihood opportunities for the local communities.

Case Study 1: The journey of Kerala from the forest to the pharmacy: A success story of an Indian state in medicinal plant research.

One of the South Indian States, Kerala provides a good case for developing countries to learn about sustainable health system development and improving its health status at low cost. The State is often praised for its good public distribution system, high political awareness, social engagement, extensive primary health care facilities, high literacy (96.9%), successful land reforms, high public participation in political process, strong labour unions and gender equity. These have also been indicated as the reasons for high health status in the state by several researchers including Ramachandran, 1997.

Kerala state was formed in 1956 by merging two regions in southern India called Malabar (part of former Madras state) and Travancore-Cochin. When Kerala state was formed, a well developed health care infrastructure was already in place and in 1957 there were 277 hospitals and 9,473 beds in the state. With steady progress the State achieved the target of 1 bed for every 1,000 people in 1971. From 1960 to 1990, the government health infrastructure further grew and reached 36,000 beds in 1986 (Kutty, 2000).

In a survey conducted in 1948, considerable advance was found to have been made in reducing diseases like malaria and filariasis, and by 1965Kerala had become the first state in India to eradicate malaria. While the health service progressed well, Kerala was lagging in the field of medical education as the first medical college was opened only in 1950 till which period the students were educated in other states. Today, Kerala has extensive, functional healthcare facilities of both

Allopathic medicine as well as TM, which has been shown in the Table 3 below (Unnikrishnan P., 2015).

Table 3: Medical Institutions, Beds & Patients Treated in Kerala during 2007

Public Sector				*Private Sector*			
S. No.	*System of Medicine*	*Institutions*	*Beds*	*S. No.*	*System of Medicine*	*Institutions*	*Beds*
1.	Allopathy	1,279	45,553	1.	Allopathy	4,825	57,071
2.	Ayurveda (TM)	871	4,020	2.	Ayurveda (TM)	4,332	5,502
3.	Homeopathy	561	1,170	3.	Homeopathy	3,226	813
	Total	**2,711**	**50,743**		**Total**	**15,094**	**114,129**

Source: Economic Review 2007, Government of Kerala.

By 2004, the infrastructural development in the state corresponded with the United States and other developed regions with around 33 beds per 10,000 population. Study of private medical institutions in Kerala shows that they constituted 34% Allopathy, 39% Ayurveda and 24.7% Homeopathy institutions. The presence of other systems such as Unani and Siddha were marginal. However, the in-patient facilities of modern medicine far outnumbered other systems as Ayurveda and homeopathy institutions were mostly dispensaries and clinics (Sankar and Deepa, 2001). Also, in terms of the number of staff, the TRM hospitals were relatively smaller compared to that of allopathic medicine.

The elements of traditional culture have therefore combined in Kerala state in a manner quite different from most other traditional communities of India. This promoted an intelligent awareness and concern regarding health, and in many ways created a receptive environment for the acceptance of the modern healthcare system. This is illustrated by the fact that till date both the indigenous and the western systems of healthcare continue in parallel, and a large part of the population makes use of both these systems selectively for different ailments and diseases.

Case Study 2: Lakadong Turmeric: A spice to help farmers triple their income.

Lakadong is a variety of turmeric that is unique to the world with the potential to change the lives of countless farmers, if its uniqueness is properly exploited. It is indigenous to the north eastern state of Meghalaya in India and is grown in a small region called Jaintia Hills. According to Meghalaya Institute of Entrepreneurship, 2018, the famous "lakadong" variety with curcumin content of about 7.4% (almost 2% higher than other varieties as proven through trials and tests by the Indian Council of Agricultural Research) is chemical free, and is highly sought after for use in the cosmetic, pharmaceutical and food industry. This makes the lakadong of particular interest to medicine and supplement manufacturers who use higher content curcumin in their products for better efficacy. This variety of turmeric has an established demand and buyers are willing to pay a premium for its quality. However, the variety and the name is being threatened by other producers who have latched onto the name to sell sub-standard or even adulterated turmeric, thereby confusing the market and lending a bad name to the original areas from where it is produced.

The antiseptic and anti-inflammatory properties of turmeric (*Curcuma longa*) have been known to Indians for centuries and with the advent of modern technology, awareness of its nutraceutical value and use as a medicine to heal, many health disorders like liver problems, digestive disorders, skin diseases and wound healing has been continuously increasing. It is commonly used as a preservative, and food colouring. Curcumin is the active ingredient of turmeric which has been shown to have a wide level of therapeutic effects.

However despite the good demand, farmers were not able to realize the full economic potential of this crop; primarily because of preponderance of small & marginal farmers, absence of focussed research, low individual volumes exacerbated by lack of organised aggregation, weak post-harvest management & market facilities, lack of universal access to information, skills & technology, unreliable price discovery, dominance of middle men, trader cartelization, weak extension, lack of assured irrigation, access to finance and most importantly, insufficient planting material in Lakadong.

In order to tackle this problem, Meghalaya launched a convergent mission of the Directorate of Horticulture, Department of Agriculture with the Government of Meghalaya termed as Mission Lakadong (2018–2023) to produce 50,000 metric tonnes of the famous Lakadong turmeric per annum in the next five years with the objective of enhancing the livelihood and income opportunities of farmers by leveraging on the uniqueness of Lakadong variety. As of now, the state produces about 16,000 metric tonnes of Lakadong turmeric per annum and the goal of the mission is to increase production with the aim of building the economy around Lakadong (Banbitlang M., 2018).

The government of Meghalaya for this reason has proposed to invest 1.5 crore USD for the promotion of Lakadong turmeric (also called golden spice) and has urged the farmers to avoid marketing of raw rhizomes and instead market value-added products like turmeric powder and turmeric oil that will fetch them a better price.

The mission also aims to build up the brand equity of Lakadong as a uniquely Meghalayan premium product (both online and offline) that would be synonymous with quality and purity so as to register the Geographical Indication for Meghalaya's Lakadong. It also focuses on creating an ecosystem that would catalyse the emergence and growth of local private enterprises in the value chain for sustained capacity building of the potential entrepreneurs.

3. NEED FOR POLICY MAKING IN THE TM SECTOR

TM has an important role to play in the society, and even the WHO supports its practice to complement MM. In order to fully achieve the objective of promoting, regulating, standardising and integration of TM in healthcare treatments, certain problems and challenges have to be overcome.

The first and foremost challenge is to document TMK to preserve, protect, and commercialize TM. The documented population-based data on the use of herbal medicinal products and IK among the younger generations are lacking, as stated by Alade *et al.* (2014). Today, the cultural survival of many indigenous communities is threatened, and some traditional systems of disseminating knowledge may already be lost. Modern lifestyles and the disruption of traditional ways of life may cause younger generations to lose interest in learning about traditional medicine. Documenting TMK may thus help preserve this knowledge for future generations and can be a vital step in facilitating research on TM safety and efficacy. In addition, documentation may assist with clinical practice, teaching and in improving public health. TMK may be misappropriated when third parties use them without the informed consent of TK holders and without equitable benefit-sharing (encompassing unauthorized use as well as misappropriation of biological resources and/or associated knowledge is referred to as "biopiracy" as stated by Karin Timmermans, 2001). However, guarding these resources has historically proven challenging due to a lack of national and international regulations and the need for law enforcement.

Documentation is necessary to obtain certain types of IP protection, which may help TMK holders to market traditional medical based products and services, facilitating investment and innovations related to TM. Documenting TMK may also be useful for defensive protection of TM that prevents third parties from improperly obtaining IP rights over TMK. The most prominent example of defensive protection has been the Traditional Knowledge Digital Library (TKDL) established by the Indian governmentto help defeat patent applications it views as inappropriate. It also restricts this data from public access in part to protect potential intellectual property rights (IPRs).

Another fundamental challenge to TM is the limited existence of research programs in the sector which is mainly confined to drug research and neglects other aspects such as production efficiency, conservation, agrotechnology and so on. Further, population demand is also lacking in the field of TRM with changes in cultural traditions and dietary habits. Gunatilleke (1984) says "…new health problems associated with the lifestyles of affluent societies are emerging in both Kerala state and Sri Lanka as a result of the cultural dislocations and the changes in lifestyles and values." This is corroborated by findings from other studies that identified that the per capita expenditure on health reduced with traditional lifestyle (Suneetha, 1998).

Despite the importance of TM for public health in many parts of the world, like the current spasm of plant and animal species extinction, as remarked by Anyinam C. (1995), the practitioners of ethnomedicine (especially herbalists and cult healers) appear to be at a greater risk of extinction than even forests and other biomes. Knowledge of the use of plants is disappearing faster than the plants themselves. The destruction of tropical forests has meant, in many parts of the tropical region, increasing disappearance of native peoples who have been living in these areas and who have accumulated a compendium of folk knowledge about the usefulness of plants for curing various diseases.

Developed countries are accruing enormous benefits from medicinal plants which they acquire from developing countries, but they are reluctant to develop benefit sharing schemes with developing countries. In a bid to entrench their position on the global arena and assert their economic dominance, developed nations have crafted international laws which protect themselves. The challenge as stated by Archibold Mposhi *et al.* (2013) for developing countries is how to protect their biological resources from being accessed freely by developed countries as well as to press for sharing of benefits derived from the use of these resources.

Documenting TMK can have far reaching consequences on intellectual property protection, commercialization and promotion of traditional medicine, regulatory submissions and interactions with collaborators. It is important that TK holders be adequately informed to safeguard their reputations and interests when interacting with third parties. Moreover, on-site communities with extensive knowledge of local environments may be used towards collaborative conservation and management. Examples of such practices can be found at the Mamirauá Sustainable Development Reserve, Brazil (Castello L., 2004; Gillingham S., 2001), in Zimbabwe (Child, B., 1996), and in the Philippines (Pomeroy RS, Carlos MB., 1997).

Exclusion of local communities from the consultation/decision processes, on the other hand, may lead to the construction of public policies devoid of historical information, and without much social resonance. Furthermore, preservation of valuable remedy of the ethnic minority groups has to be taken up to minimise the current distrust between modern and traditional doctors and to achieve the objective of regulation, standardisation and cooperation.

3.1 Policies Required for Protecting and Popularizing Traditional Medicinal Knowledge

In the era where S&T is developing rapidly, new knowledge is being discovered, new inventions are being made, and researchers are becoming outmoded in a few years time, the increasing emphasis is now on science education. The displacement of one paradigm by another as a guiding principle of research in the scientific community is known as a 'paradigm shift' or 'scientific revolution'. However, science education as it is practiced in the world today often entails little more than training in the concepts and principles of existing knowledge while simply passing on bits of scientific information.

TMK these days is experiencing increased attention for global health care demand of developing countries. So much so that in some Asian and African countries, upto 80% population relies on traditional medicine for their primary health care needs. Therefore, it is necessary that the Government should encourage the use of traditional medicines and herbal remedies, not only in rural areas where there are few or no hospitals and medical clinics but also elsewhere in the country. Also, the political and scientific leadership of the developing countries should promote TMK as a vehicle for the onward journey to progress and believe in the capacity of TM especially for income generation.

Indigenous people, perspectives and knowledge have contributed substantially to the development of S&T throughout history and will continue to do so for generations to come! TM is a wonderful alternative medical system but has its own limitations. Although, nothing can compare to MM during an acute emergency like heart attacks, stroke, paralysis, fracture, when it comes to preventing disease and treating chronic illnesses, MM falls short of expectations. TM thus remains to be a system that has stood the test of time, based on the concept of perfect health and has come to prominence in the western world during the last decade.

Developing countries are urgently required to make efforts to protect TMK through national legislation and international agreements. Some of the other policies that may be adopted by the government entities of the developing countries to popularize TMK could be:

- Formulate national policy and regulation for the proper use of TM and its integration into national health care systems in line with the provisions of the WHO strategies on TM;
- Promote therapeutically sound use of TM by providers and consumers;
- Establish standards for safety and quality of products;
- Train the younger generation about age-old practices through their school curriculum to preserve the country's IK;
- Provide adequate funding for bio-prospecting and TM research;

- Encourage and assist TK holders to first protect and then share their knowledge in public platforms;
- Create a stronger evidence base on the safety, efficacy and quality of the TM products and practices;
- Ensure availability and affordability of TM especially the essential herbal medicines;
- Cultivate and conserve medicinal plants to ensure their sustainable use;
- Develop national policies on the evaluation and regulation of TM practices while ensuring effective regulatory oversight;
- Make TMK a prime vehicle of national development by fostering, promoting and sustaining the cultivation of indigenous and scientific research in all aspects, pure, applied and educational;
- Encourage indigenous science scholar recruitment incorporating indigenous perspective to promote culturally inclusive, scientific approaches; and
- Promote integration of the use between modern drugs and traditional medicines effectively and safely, in disease prevention and treatment.

4. CONCLUSION

Traditional medicine occupies a particularly important place in health care sector, in the world and thus should be effectively developed through an integrated environmental, national and international framework set up by the Government. By exploring more about natural environment and focusing on research related to the use of plants, TMK can then be used to fight poverty and improving welfare in the developing countries.

The Governments should establish the necessary institutional and financial support to promote the potential role of herbal medicine in primary health care delivery. Priority should be given to the development of herbal medicine by means of the following measures: (a) inventorying and documenting the various medicinal plants and herbs which are used to treat common diseases in each region of the country; (b) establishing local botanical gardens for the preservation of essential medicinal herbal plants in different parts of each country, in order to ensure a sustainable supply of safe, effective and affordable medicinal herbs; and (c) setting up testing laboratories with adequate facilities for the assessment of the efficacy of medicinal herbs, and establishing dosage norms for the proper administration.

Government budgets, together with direct investment from private sector companies, would be required to implement the measure. Further, an increased access to quality traditional health services and remedies can also be expected through the implementation of the measure.

ACKNOWLEDGEMENT

I would like to express my sincere gratitude to Dr. Amitava Bandopadhyay, Director General, NAM S&T Centre and Mr. M. Bandyopadhyay, Senior Advisor for their thoughtful insights on the content of the paper. I hope that through this paper, the developing countries will give due relevance to policy making for popularizing the indigenous medicinal plants (an invaluable asset) in their country and use it as a means for improving their income, entrepreneurial potential and their overall lifestyle.

REFERENCES

[1] Abiose, Adeleja (2006). Nigeria boosts research into traditional medicine. Science and Development Network. www.scidev.net/en/agriculture-and-environment/news (accessed on 8 December 2019).

[2] Alade et al. (2014). Indigenous knowledge of HM among adolescents in Amassoma, Bayelsa State, Nigeria G.O. Global Journal of Health Sciences, 8(1), 2016, ISSN 1916-9736.

[3] Andrea Rinaldi , Priya Shetty (30 June 2015). Traditional medicine for modern times: Facts and figures. SciDev.net. https://www.scidev.net/global/medicine/feature/traditional-medicine-modern-times-facts-figures.html (accessed on 13 November 2019).

[4] Anyaoku, E.N., Orizu, O.E.N. and Eneh, E.A. (2015). Collection and Preservation of Traditional Medical Knowledge: Roles for Medical Libraries in Nigeria. Journal of Library and Information Sciences, 3(1), 33–43.

[5] Anyinam, C. (1995). Ecology and Ethnomedicine: Exploring Links between Current Environmental Crisis and Indigenous Medical Practices. Social Science & Medicine, 40(3): 321–329. doi: 10.1016/0277-9536(94)E0098-D.

[6] Avendaño, Carmen and Menéndez, J. Carlos (2008). Medicinal Chemistry of Anticancer Drugs, Elsevier, ISBN:9780444528247.

[7] Banbitlang Marbaninang (2018). Mission to promote famed Lakadong turmeric, The Telegraph (Online Edition). https://www.telegraphindia.

com/states/north-east/mission-to-promote-famed-lakadong-turmeric/cid/1447255(accessed on 23 November 2019).

[8] Barrett, B. et al. (1999). Assessing the risks and benefits of herbal medicine: An overview of scientific evidence, Altern. Ther Health Med., 40–9.

[9] Bernard, H.K. Luk (1989). Population of S&T. What informal and non-informal education can do? Faculty of Education, University of Hong Kong - UNESCO, 1989, 210 p.

[10] Castello, L. (2004). A Method to Count Pirarucu Arapaima gigas: Fishers, Assessment and Management. North American J. Fish Managemt, 24:379–389. doi: 10.1577/M02-024.1.

[11] Child, B. (1996). The practice and principles of community-based wildlife management in Zimbabwe: the CAMPFIRE programme. Biodiversity and Conservation 5(3):369–398. doi: 10.1007/BF00051780.

[12] Daily, G.C. and Ehrlich, P.R. (1996). Global change and human susceptibility to diseases. Annu Rev Energy Environ., 21:125–44. doi: 10.1146/annurev.energy.21.1.125.

[13] Dan, Bensky *et al.* (2004). Chinese Herbal Medicine Materia Medica, 3rd edition, 734 (Eastland Press).

[14] Daniel, S. Fabricant and Norman, R. Farnsworth (2001). The Value of Plants Used in Traditional Medicine for Drug Discovery, Environmental Health Perspectives, pp. 69–75.

[15] Directive 2004/24/EC, 2004 O.J. (L 136) 85, amending, as regards traditional herbal medicinal products, Directive 2001/83/EC. http://eur-lex.europa.eu/LexUriServ/LexUriServ.do?uri=OJ:L:2004:136:0085:0090:EN:PDF (accessed on 19 September 2019).

[16] Economic Review (2007). Government of Kerala. http://www.spb.kerala.gov.in/images/pdf/ER1959_2009/2007.pdf(accessed on 23 October 2019).

[17] Fifty Sixth World Health Assembly, Agenda item 14.10, Traditional Medicine, WHA56.3, May 28, 2003. http://whqlibdoc.who.int/wha/2003/WHA56_31.pdf (accessed on 14 October 2019).

[18] Gillingham, S. (2001). Social Organization and Participatory Resource Management in Brazilian Ribeirinho Communities: A Case Study of the Mamirauá Sustainable Development Reserve, Amazonas. Society & Natural Resources, 14(9): 803–814.doi: 0.1080/089419201753210611.

[19] Jesse, Popp (January 3, 2018). How indigenous knowledge advances modern science and technology, The conversation.com. https://theconversation.com/how-indi-know-89351 (accessed on 9 September 2019).

[20] Karin, Timmermans, TRIPS, CBD and Traditional Medicines: Concepts: Concepts and Questions, Report of an ASEAN Workshop on the TRIPS Agreement and Traditional Medicine (February 13–15, 2001), available at http://apps.who.int/medicinedocs/en/d/Jh2996e/6.4.html (accessed on 23 October 2019).

[21] Kutty, V.R. (2000). Historical Analysis of the Development of Health Care Facilities in Kerala State, India, Health Policy and Planning, 15, pp. 103–9.

[22] Meghalaya Institute of Entrepreneurship (2018). Mission Lakadong, 2018–2023, pp. 13–24.

[23] Mposhi, A., Manyeruke, C. and Hamauswa, S. (2013). The Importance of Patenting Traditional Medicines in Africa: the case of Zimbabwe. International Journal of Humanities and Social Science, 3(2) (Special Issue, January 2013)].

[24] Pomeroy, R.S. and Carlos, M.B. (1997). Community-based coastal resource management in the Philippines: A review and evaluation of programs and projects, 1984–1994. Marine Policy, 21(5): 445–464. doi: 10.1016/S0308-597X(97)00016-X.

[25] Ramachandran, V.K. (1997). Kerala's Development Achievements in Dreze Jean and Amartya Sen eds., Indian Development—Selected Regional Perspectives, Delhi: Oxford University Press, 1997.

[26] Ryan, Abbott (2014). Documenting traditional medicinal knowledge. World Intellectual Property Organisation (WIPO). https://papers.ssrn.com/sol3/papers.cfm?abstract_id=2406649 (accessed on 17 October 2019).

[27] Ryan, B. Abbott *et al.* (2011). Medical Student Attitudes Toward Complementary, Alternative, and Integrative Medicine, Evidence Based Complementary and Alternative Medicine, Vol. 2011, Article ID 985243, 14 pages.

[28] Sankar, Deepa, (2001). The Role of Traditional and Alternative Systems in Providing Health Care Options: Evidence from Kerala, Institute of Economic Growth, Delhi Discussion Papers, Institute of Economic Growth, Delhi, India.https://www.worldcat.org/title/role-of-traditional-and-alternative-systems-in-providing-health-care-options-evidence-from-kerala/oclc/56097116 (accessed on 23 October 2019).

[29] Suneetha, M.S. (1998). Demand for and Valuation of Medicinal Plants and Products in Western Ghats of Kerala—A Resource Economics Study. Bangalore: University of Agricultural Sciences (Unpublished research thesis).

[30] Unnikrishnan, Payyappallimana (2015). Traditional Medicine in Health System Development, Yokohama Journal of Social Sciences, 15(3).

[31] Vagelos, P.R. (1991). Are prescription drug prices high? Science, 1080–4.

www.megagriculture.gov.in › PUBLIC › file Upload › download (accessed on 4 October 2019).

APPENDIX

Indian Systems of Medicines (AYUSH)

Ayurveda: This system of medicine originated in Vedic period and the medical knowledge in Atharveda generally developed into the science of Ayurveda. The diagnosis in Ayurveda involves questioning, examination of pulse, tongue, urine etc. This also involved inference treatment such as preventive measures and curative measures that include intake of drugs and medicines. Ayurveda also makes use of the medicinal properties of the plants.

Yoga: A way of life propounded by Yajyavalkya and later systematized by Patanjali. The history of yoga is about 2500 years when Patanjali provided it a systematic form which consists of eight components. This yoga helps in improving physical, mental and social-wellbeing and provides relief from stress and strain, increases resistance power of the body, etc.

Naturopathy: Like yoga, it is a way of life and believes in the practice of applying the simple laws of nature to life. It is commonly known for drugless treatment of disease and an organized way of life is the key to health energy and happiness. It lays special emphasis on eating and living habits, adoption of purificatory measures, use of hydrotherapy, cold packs, mud packs, massage, etc.

Unani: This system of medicine originated in Greece but was brought to India by Arabs and Persians. However, this system of medicine has a long impressive record in India. It also adopts holistic approach. But main types of treatment are regimen therapy, diet therapy, pharmacy-therapy and surgery.

Sidha: The most ancient Indian system of medicine practiced in India. The basic principles and doctrines of Sidha system are similar to those of Ayurveda. Both Ayurveda and Sidha are holistic system of medicines dealing with preventive, primitive and curative concepts of health and are holistic systems of medicine covering, body, mind, soul and phenomena of nature. But Sidha was primarily practiced in Southern part of India.

Homeopathy: This system of medicine was discovered by German physician Dr. Samuel Hahnemann and cures natural disease by administering drugs which have been experimentally proved to possess the power of producing similar artificial symptoms on healthy human beings. It has also wide acceptability because of its holistic and intrinsic values, low cost medicines and absence of any side effects of the drug.

TEHRAN RESOLUTION

Policy Making in Popularisation of Science, Technology and Innovation (STI) in Developing Countries

TEHRAN RESOLUTION

on

Policy Making in Popularisation of Science, Technology and Innovation (STI) in Developing Countries

We, the Delegates of the International Workshop on "Policy Making in Popularisation of Science, Technology and Innovation (STI)", organized by the Centre for Science & Technology of the Non-Aligned and other Developing Countries (NAM S&T Centre) in partnership with the National Research Institute for Science Policy (NRISP), Iran at Tehran during 17–19 June 2019.

Comprising researchers and scientists in academic and S&T institutions, and policy makers in the government departments and ministries of the developing countries, viz. Bhutan, Cambodia, Colombia, Cuba, India, Indonesia, Iran, Malaysia, Mauritius, Myanmar, Nigeria, Pakistan, Palestine, Sri Lanka, Togo and Zimbabwe.

Expressing Gratitude to the organizers of the International Workshop, the NAM S&T Centre, New Delhi and the National Research Institute for Science Policy (NRISP), Tehran.

Considering the efforts made by the developing countries to make science and technology an essential part of their culture, which is of paramount importance to the overall development of the nation as popularization of science, technology and innovation is synonymous with sustainable development.

Recognising that STI policies that are well integrated into national development strategies and combined with institutional and organizational changes can help raise productivity, improve competitiveness, improve creative thinking abilities, science and technology literacy, public engagement and public monitoring and assessment.

Committing to identify key challenges to promote sustainable innovation in developing countries; to support integration of STI in national development strategies; to build up and strengthen the existing STI policy making capacities of the developing nations; setting of agenda by using national educational programs to execute policy formulation by collaborating with appropriate stakeholders and communicating with the policy-makers and civil society organizations; and to facilitate STI for a sustainable socio-economic development of the developing countries.

Deliberating on the importance of formulating and implementing appropriate policies on STI popularization for inclusive and sustainable development.

Noting the necessity for new approaches that embed the policies on STI popularization and capacity-building as crucial components of national development plans.

Noting also that the STI popularization must be aligned to address various aspects of sustainable development.

Contemplating that the popularisation of STI help people to get familiarized with the utility and benefits of science and technology for improving their social and economic life, popularisation of STI should be an essential and prime component in the formulation and implementation of STI policy in the developing countries.

Taking Note of the importance for STI development policies, programmes and projects to address various aspects of the STI popularization, particularly the policy-making and policy implementation processes.

Taking into Consideration that the successful policies on popularizing of STI at the national level is facilitated by, among other things, governmental, non-governmental and civil sectors, to create policy environments that enable education and research institutions, businesses and industry to innovate, invest and transform science, technology` and innovation into employment, educational and economic growth incorporating all interrelated elements.

Unanimously Recommend that governments, policy makers, STI institutes, researchers, educationists and scientists are encouraged to consider taking the following actions:

- To closely link the policies on STI popularization and strategies of national development planning, by prominently featuring capacity-building in science, technology and innovation policies.

- To undertake systemic research, for foresight programmes, on new trends in STI policies and their impact on popularization of science, technology and innovation.
- To identify potential education gaps in STI popularization and address such gaps with a policy mix, including the promotion of STI education and vocational trainings through informal/non-formal educational systems and settings.
- To raise awareness among policy-makers regarding the importance of policies on STI popularization, with special attention being placed on new trends in STI that can offer novel possibilities for developing countries.
- To encourage structured debate among all stakeholders, including representatives of governments, science communities, industry, civil societies and the private sectors, towards creating a shared understanding of issues related to STI policies.
- To fill the ongoing and persistent gap in public understanding of STI, by encouraging, mentoring and supporting all efforts to attract and retain experts and others into the field.
- To encourage scientists, researchers, innovators, policy-makers and other related professionals in developing countries to involve themselves in science popularisation and communication as a part of their social responsibility.
- To carry out research in different modes for the popularisation of STI.
- To ensure that policies for STI popularisation must address all the three dimensions of sustainable development i.e. social, economical and environmental.
- To allocate/increase national funding for research and development in STI popularization and support with appropriate resources.
- To facilitate networking and partnerships among all the members of NAM and other developing countries at the national and international levels, to strengthen collaboration on popularization of STI.
- To provide a specific forum for sharing not only success stories and best practices but also failures and key challenges and learnings from the results of foresight programmes, successful local STI popularization models, case studies and experiences on the use of STI popularization strategies.

- To encourage media institutions, news agencies and journalists to pay more attention towards STI popularization and create appropriate infrastructures to raise public awareness and engagement with STI, and communicate more effectively with scholars, scientists, administrators and others in the education and research institutes.
- To consider the essential role of science centers/museums/science parks/libraries/events/shows/games/etc. in promoting STI popularization.

Thus Resolved and Adopted on the 19th Day of June, 2019 in Tehran, Iran.

Author Index